The Preservation of Species

This book was written under the auspices of the
Center for Philosophy and Public Policy,
University of Maryland

The Preservation of Species

The Value of Biological Diversity

EDITED BY

BRYAN G. NORTON

PRINCETON UNIVERSITY PRESS
PRINCETON, NEW JERSEY

Publication of this book has been aided by a grant from
the Harold W. McGraw, Jr. Fund of Princeton University Press

Library of Congress Cataloging in Publication Data will be
found on the last printed page of this book

First Princeton Paperback printing, 1988

This book has been composed in Linotron Sabon and Gill Sans

Clothbound editions of Princeton University Press books
are printed on acid-free paper, and binding materials are
chosen for strength and durability. Paperbacks, although satisfactory
for personal collections, are not usually suitable for library rebinding

Printed in the United States of America by Princeton University Press
Princeton, New Jersey

For R. T.

Contents

Contents

Preface

The Center for Philosophy and Public Policy was established in 1976 at the University of Maryland in College Park to conduct research into the values and concepts that underlie public policy. Most other research into public policy is empirical: it assesses costs, describes constituencies, and makes predictions. The Center's research is conceptual and normative. It investigates the structure of arguments and the nature of values relevant to the formation, justification, and criticism of public policy. The results of its research are disseminated through workshops, conferences, teaching materials, the Center's newsletter, and books like this one.

The essays presented here grew from the deliberations of a working group assembled by the Center. The group consisted of biological scientists, social scientists, attorneys, resource managers from both the private and public sector, and philosophers. The goals of the project were, first, to examine and evaluate the various reasons that have been and could be given for preserving nonhuman species from extinction and, second, to grapple with the difficult problem of priorities—what should be done when financial and other resources are insufficient to save all species? It was assumed at the outset that a rational answer to the priorities question could only be stated in the context of a careful examination of the reasons for preserving species.

The interdisciplinary group assembled was chosen in order to stimulate new thinking about a recalcitrant subject, rather than to represent standard, accepted positions. While we tried to ensure that all necessary technical expertise was available in the group, we tended to choose representatives of the various professions and disciplines according to their reputation for creative thinking over a range of problem areas, in preference to narrow expertise on the problem at hand. We encouraged members to think broadly, abstractly, and "philosophically" about the applications of knowledge from the field of their expertise. We first gathered in Washington, D.C., on December 4, 1981, to discuss the scope and objectives of the group, and each member accepted responsibility for addressing the endangered species problem from a particular point of view. We returned to Washington twice more to discuss each other's first drafts. These discussions shed light from diverse disciplines upon papers that, necessarily, were initially written from one disciplinary point of view. The spirit of common concern and cooperation displayed at those meetings is responsible for the interdisciplinary flavor of the essays resulting from subsequent rounds of rewritings.

The purpose of this book is not to present new data or new scientific theories. Nor is it to inform conservation biologists acting as hands-on managers of populations or ecosystems. It is, rather, to analyze and interpret existing scientific data, presenting it in conjunction with analyses of the problem and its possible solutions. As such, it is addressed to those who are interested in broad policy issues regarding the preservation of species. For references geared more to the needs of biologists faced with day-to-day decisions affecting the persistence of a particular species, see Chapter 9 and the Bibliography.

Our audience should be—besides interested laypersons—legislators, activists, policy-makers, and higher-level managers, people often not trained as biologists. The discussions here bear upon general problems of preservation and are designed to provide a context in which managers can plan a strategy. Often the guidance afforded will help nonspecialists to decide when specialized expertise must be sought. As such, this book should not be seen as a competitor with texts in conservation biology, but rather as providing an overview of the problems with which they deal. We know of no such general guide and believe the book makes a unique contribution in this respect.

We have tried to address the problem of endangered species and its concomitant effect, a reduction in biological diversity, in its most general terms. To address a problem of this magnitude, however, requires decisions affecting emphases. In particular, there is less discussion of biological, attitudinal, and political matters from an international perspective than would be ideal. Most decisions are made within a polity, and it seemed better to try to achieve a solid understanding of factors affecting decisions in U.S. policy than to attempt global coverage. We believe, however, that the information presented here can inform decisions internationally.

Likewise, more examples and generalizations are drawn from the animal than the plant kingdom, often because more data exist to support them. Finally, there is little discussion here of gene banks and methods of preserving seeds and germplasm. That they represent essential complements to the *in situ* approaches emphasized here is unquestioned. But, since the advocates of these alternatives insist that they are not designed to replace preservation in the wild, we have not attempted to duplicate their efforts. Important works on this subject are included in the Bibliography.

No special efforts were made to forge a consensus and many points of disagreement remain. Areas of consensus and important areas of disagreement are summarized in the Epilogue. This Epilogue, composed by the Editor, was approved by all members of the group, although there is no reason to believe that all members would choose the same points of emphasis.

We must acknowledge the help of many institutions and individuals, especially the National Science Foundation and the National Endowment for the Humanities for the funding through the EVIST program that made the project possible (grants IPS-80-24258 and IPS-80-24358), as well as the program coordinators, Rachelle Hollander and Eric Juengst, who were helpful throughout. Mark Sagoff, who conceived the project and wrote the proposal, offered expert guidance and perspective, while also supplying humor as an important ingredient. Peter G. Brown, Director of the Center while the grant was shepherded through the university and agency processes, contributed greatly to the project's launching. Above all, the editor wishes to thank Henry Shue who, as Director of the Center during the execution of the project, offered continuous support and good advice.

Thanks are also extended to the working group members who took time from their busy schedules to write papers, attend meetings, and respond to our many and sometimes onerous requests for more work. Others, too numerous to mention, spent time discussing the project with the principal investigator and suggesting prospects for working group membership. Their help was invaluable.

The computer time for the production of this book was provided by the Computer Science Center of the University of Maryland, and we thank them for their ongoing assistance in its production.

Finally, we thank the other members of the Center's research staff for their valuable suggestions and help throughout the project and the Center's support staff for their indispensable assistance in the production of materials and for astoundingly good nature against impossible odds. Claudia Mills improved our writing, especially by shortening papers without sacrificing their arguments. Elizabeth Cahoon and Rachel Sailer administered the grant without a hitch and freed our time to concentrate on research. Carroll Linkins did all of the correspondence, filing, and much of the typing, while keeping her sense of humor amidst chaos. Louise Collins, Lori Owen, Robin Sheets, and Virginia Smith all pitched in, in countless ways. The successful completion of this volume owes much to many hands, and their labor is gratefully acknowledged.

Bryan Norton
February 1985

The Preservation of Species

Introduction

Since the emergence of the human species something over 300,000 years ago, it has grown in population and spread across the globe. This growth began slowly, but has recently accelerated rapidly. For much of history, human life was precarious; while comparatively intelligent, we lacked physical strength and other protections necessary to secure life easily. But in recent centuries, through our technology, we have gained a level of dominance probably never achieved by one species over the others in the history of life on earth. Human shelters now provide outposts on every area of the globe as we have expanded the range of our habitat. Highly organized social structures, techniques of communication, and methods of transportation allow, although not without severe problems, huge concentrations of people in metropolitan areas. Highly mechanized agricultural and extractive techniques make possible food and resource production on a scale undreamed of only decades ago.

Until recently, these developments were described with little skepticism as "human progress." But this progress against the elements and against competitors from other species is now undergoing reassessment. A dramatic trend toward simplification and alteration of habitats, as well as alarming projections of species extinctions, has frightened thoughtful observers. If projections that up to one quarter of living species may be lost in the next twenty-five years prove even approximately accurate, the biological world faces an upheaval of staggering proportions.

Two extreme strategies suggest themselves. One strategy is to press forward with the task of technologizing nature. On this strategy, the human species would strive to increase its independence from nature and natural systems by continuing to create artificial environments and replace the services now derived from natural ecosystems with services derived from technological sources. A second extreme strategy would be to retreat from technological progress, reverse the trend toward manmade environments, and try to save as much of as many remaining natural systems as possible.

Neither extreme strategy appears feasible, but exploding population growth, rising expectations of material well-being, and seemingly autonomous forces driving technological growth seem certain to push human society in the direction of the first, technologically oriented strategy. But whichever strategy is chosen from the wide range of options available, a major increase in the rate of species extinctions will certainly occur and have to be dealt with in some way. Events have already progressed too far to avoid it entirely. Choices now being made affect the magnitude of these extinctions, but choices already made have determined that they will occur. This major biological event, our reactions to it, and the strategies we choose to deal with it constitute the problem of endangered species.

What kind of problem is it? One's first temptation may be to say that it is a very complex scientific problem, a question of how to understand, predict, and measure the phenomena of extinction and endangerment of species and their effects on the biological world. But the endangered species problem goes beyond purely scientific questions—at issue is also what actions should be taken. And problems demanding action prompt concern about motives, goals, and objectives, thereby raising questions of values as well.

Questions of values fall within the realm of philosophy. But the endangered species problem is also not one of developing abstractly stated goals. Rather, we must perceive the scientific issues within a larger context of competing social needs. And thus the endangered species problem must ultimately be addressed from many different perspectives. It is essential to know how many species are being lost and why, what effect these losses have on remaining species and on their habitats, the present and expected effects of these changes on human lives, and what steps could be taken to alter the trends simplifying the biological world. But knowledge of these biological facts and relationships will prove useful only if it is combined with an understanding of the socioeconomic and cultural factors that have set in motion the biological trends. And the humanities—value studies—are essential to compare what is happening with what might and should be. This book attempts, for the first time, an integrated approach to the endangered species problem.

We have divided the volume into three sections. Part I explores the dimensions of the problem. Chapter 1 concentrates on its global scope and pervasive character. Chapter 2 compares current and projected extinction rates with major extinction events of the past in an effort to understand the magnitude of the consequences the human race is now facing. Chapter 3 explores the problem in the context of socioeconomic and cultural factors.

Part II examines the values and objectives to which policy analysts must ultimately appeal in order to motivate a reversal of the trend toward biological simplification. Chapters 4 and 5, one by an economist and one by a philosopher, emphasize the utilitarian values humans derive from species diversity. The next three chapters examine reasons for preserving species based upon, respectively, the intrinsic value of other species, aesthetic values, and the intrinsic value of human knowledge of natural objects. These reasons are not, of course, exclusive—nonhuman species may be valuable in all of these ways. The writers of the essays attempt to assess, however, the comparative strengths of the motives deriving from each source of values.

Part III contains guidelines designed to help agencies and individuals charged with protecting biological diversity. Chapter 9 draws upon available ecological knowledge; while far less knowledge is available than would be desirable, it is essential that managers base their plans and projects on the best scientific information they can acquire. Chapter 10, recognizing that full agreement does not exist either on the values and goals motivating preservation activities or on many crucial scientific points, offers advice to resource managers who nonetheless must act, unable to afford the luxury of waiting upon scientific and value consensuses. The final chapter in this section addresses strategic questions concerning how to build positive incentives for private landowners into preservation efforts, in the hope of relying less upon restrictions.

Finally, the Epilogue attempts to summarize the agreements, disagreements, and unknowns that emerged from our writings and discussions. We hope that it will be valuable in locating areas of consensus and conflict across disciplines, thereby focusing future research and discussion on open and important questions.

PART I

The Problem

Introduction to Part I

What is the scope of the endangered species problem? What are its exact dimensions? What does the problem mean to ordinary people with concerns of making a living and raising a family in what remains of the twentieth century?

Thomas Lovejoy places the problem in global perspective, as it arises in both developed nations such as the United States and in the developing nations of the tropics. Affluent countries with well-developed scientific communities, with vocal environmental movements, and with legal institutions designed to address the problem of decreasing biological diversity still find the problem a recalcitrant one. In the United States, each threatened species could, conceivably, be given individual treatment. And yet the destruction of habitat continues as economic growth places more and more land under intense use for human purposes. Even species that are protected (many are not, because of institutional shortcomings) seem doomed to exist in limited ranges requiring human management in controlled preserves. Lovejoy expresses his concern that biological diversity will be lost through "incremental decision-making": the economic costs of preserving any individual species seem to outweigh the benefits, but the overall effect of negative decisions regarding preservation of individual species will be a disastrous simplification of ecosystems. The unflagging demand for increased human goods and services seems to mean that the most sincere and diligent efforts will only delay the inevitable loss of species.

In the developing nations of the tropics, deforestation continues and accelerates, with entire habitats destroyed in days or months. There is no time to show special concern for any but the best-known and most magnificent species. In many cases, fewer than half of the threatened species are even identified and named; nothing is known of their life histories and habitats. Lovejoy argues that such wholesale alteration of ecosystems is disastrous not only in biological, but also in human terms. He is convinced that failure to protect biological diversity is closely connected with failure to maintain a reasonable standard of living for human inhabitants. The vicious circle of poverty, of overexploitation of natural resources, and of degradation of water and soil must be broken if the unhappy results now observed in nations such as Haiti are not to be repeated throughout the tropics.

Best estimates project that, if present trends continue, 20 to 25 percent of existing species will be destroyed in the next quarter of a century. But do we really understand what such predictions mean? Most people today live in urban environments, seldom encountering wildlife except in controlled settings like zoos and parks. Otherwise, their interactions with nonhuman species are limited to house pets and garden plants and to infestations of pests such as rodents and roaches. Geerat Vermeij surveys the literature on actual current species extinctions and assesses what predicted future extinction rates would mean, drawing upon available knowledge of major extinction events of past epochs. The predicted rates of extinction are unprecedented in paleontological history, and once major changes in diversity begin, scientific evidence suggests that other species dependent upon lost species will be extinguished in turn. Consequently, a downward spiral in diversity is likely to continue for centuries and perhaps millennia. Species remaining will be skewed toward opportunism in life habits, and opportunistic species are generally less compatible with human goals than are more highly specialized species. Regeneration of species from limited stocks remaining after a major extinction event is possible, but occurs very slowly, taking thousands of years. An assessment of the paleontological literature is, then, sobering indeed.

Stephen Kellert assesses the problem of extinction in its social context. He concludes that the problem is fundamentally rooted in social and political forces. Therefore, attempts to address the issues only in scientific and wildlife management terms are sure to fail. A more profound effort, to understand and change human attitudes toward other species, is essential. In the short term, this entails exploring, analyzing, and publicizing ways in which human life and happiness in all cultures depend upon the preservation of other species and natural habitats. In the longer term, it means developing a closer personal and spiritual sense of relatedness to the nonhuman world.

The problem of endangered species, then, must be understood on several levels. First, it must be understood globally. In the United States, a wealthy nation still experiencing population growth, increased use of technology, and accompanying fragmentation of habitats, the problem seems one of identifying individual endangered species and finding means to minimize the impacts of economic growth on their habitats. In the nonindustrialized world, where economic development is only beginning and many individuals live in great poverty, much more radical changes in the environment can be expected. Can those changes be made while minimizing impacts on ecological systems? It is certainly true that all peoples, in developed and developing nations alike, will be affected by losses in worldwide biological diversity. One of the most difficult components of the challenge we face is thus a political one: what forms of international cooperation can be arranged so that both rich and poor nations can address the global, long-term problems effectively?

Second, scientific data imply that the effects of current losses of species will reverberate through ecosystems for decades, centuries, and millennia. But so little is actually known about how species interact in ecosystems, about dependency relations, or about how ecosystems recover from disturbances, that actions required now to avoid future disasters often must be undertaken without sufficient knowledge to make considered choices. Only a scientific effort of unprecedented proportions could provide the information necessary to make wise decisions affecting the future.

But a scientific effort cannot be successful by itself. As long as decision-makers treat other species and natural ecosystems merely as sources of raw materials to be exploited for short-term gain, as long as human perceptions and attitudes set people apart from nature, and as long as the worth of a life form is measured in terms of its phylogenetic proximity to the human species or in terms of its short-term economic value, no amount of scientific knowledge will solve the problem of endangered species. The problem is, on the deepest level, not scientific but perceptual and attitudinal.

I

Species Leave the Ark
One by One

THOMAS E. LOVEJOY

Endangered species are usually perceived as representing no more than a small number of highly individual and esoteric situations—such as the mysterious small tree from the Alatamaha River Valley of Georgia discovered by John and William Bartram in 1765, named in gratitude for Benjamin Franklin, and found only a few times in nature and not since 1803.[1] This perception of disappearing individual species as opposed to a process of biotic impoverishment is a natural one because people identify readily with individual plants and animals, or individual species, even if the organism is only recently familiar to the general public.

In recent years, however, awareness has grown that the problem is actually a matter of considerably greater magnitude than heretofore realized. To a large degree, this is a consequence of the accelerating rates of tropical deforestation.[2]

I. How Many Are There?

Estimates of numbers of endangered species were first restricted to warm-blooded vertebrates, because birds and mammals are among the more noticed elements of the biota. Today, endangered species lists prepared by the U.S. Department of the Interior and by the Species Conservation Monitoring Unit of the International Union for the Conservation of Nature and Natural Resources (IUCN) attempt to encompass all forms of life. This is a very difficult task because most species of plants and animals are still completely unknown to science and there is no information about their status or biology.

Estimates of the magnitude of the problem are usually put in terms of projections (which are not to be equated with predictions) of the number of extinctions that might be expected by the end of the century. These projections assume no major changes in various trends such as population and deforestation. The number of extinctions projected ranges from hundreds of thousands to over a million.

In part because the size of the numbers is startling, there has been some criticism of the figures.[3] Some of the criticism has been justified. In some instances there has been no explanation of how the estimate was obtained. In others, estimates have been made in terms of rates, implying a greater precision than is possible. Further, the meaning of a rate in terms of cumulative numbers has often been ignored. The "one extinction per minute" used by some authors is equivalent to 525,600 extinctions per year (an unlikely or impossible number).[4]

Other criticism has tended to view the large numbers as the outpouring of overwrought biological Cassandras and has asked how they could be so very much greater than historical figures.[5] This criticism ignores basic biology in trying to compare the historical figures for mammalian extinctions with the recent projections which encompass all forms of life. Historical figures for mammalian extinctions are indeed small compared to the projections but they are only meaningful compared to the total number of possible mammalian extinctions (i.e., 4,500, the total number of extant mammal species), or as an indication of accelerating extinction rates.

That there is considerable spread in the estimates is really not surprising, given the difficulties in getting precise information. Just to begin, the size of the biota is only known to an order of magnitude. Estimates of the complete array of life forms on our planet vary from three to ten million, and occasionally go higher.[6] Only about 1.5 million species have actually been described by science and, even then, their full geographic range and ecological peculiarities have not been recorded. In addition to the problem of not knowing how many species there are, what they are, and where they are, there is the additional problem that there exists no accurate estimate of the extent of the major transformations in the biological landscape such as tropical deforestation. Consequently, a considerable variation in estimates can be expected.

What is important is that every effort to estimate extinction rates has produced a *large* number. This is consistent with what can be observed of human-wrought changes in the biology of the planet. The tropical forests probably hold half or more of all species of plants and animals. Most of the species of primary tropical forests do not occur in second growth as is much more often the case in temperate zones. Nor in most instances are tropical forest plant species able to persist for long periods of time (decades and even centuries) in soil seed banks the way most temperate forest species can. Such characteristics explain why New England's forests have recovered from extensive cutting in the nineteenth century without major loss of diversity, but tropical forests will not do so.

The great numbers of species in tropical forests usually occur at very low densities (few individuals per unit area) and often they are very restricted in distribution. Large numbers of species will almost inevitably be lost from tropical deforestation. The situation is close to the end in the forests of Madagascar where 90 percent of the species are endemic (occur nowhere else), and the Atlantic forest region of Brazil (also high in endemism); the latter is reduced to less than 2 percent of its original extent. Heroic and major last-minute efforts may yet save a good portion of the highly endemic floras and faunas in such areas. Yet nobody expects tropical deforestation to cease tomorrow. The relation between area of tropical forest destroyed and the number of possible extinctions was the basis for the projections in the Global 2000 Report.[7]

Besides the urgency implied by events in the tropics, there is also a growing awareness of the implications of habitat fragmentation for potential extinctions. Habitat fragments have differing ecological dynamics from those they had when once part of continuous wilderness; this change leads the remnant fragments to lose species after isolation.[8] Isolated habitat fragments are almost always subject to this species loss process. Little is known of the process and what forces drive it. Nor is it known how ordered the species loss really is; if there is a predictable order in which species vanish from habitat fragments, those early in the order are much more prone to extinction because they are likely to be lost from every habitat fragment. The process of species loss from fragments of habitats basically calls into question whether many areas set aside to protect particular forms of life, and which are often isolated habitat remnants, will, in fact, actually do so. The activity of civilization has tended to fragment natural landscapes (when not entirely obliterating them), so that the terrestrial portion of the globe has become a biological fretwork each portion of which is going through the process of species loss.

At the same time, people are busy changing the chemistry, if not the physics, of the planet. The recent increase in acid precipitation from burning high-sulfur fuels is a prominent example.[9] The origin of acid rain is not fully understood nor are the effects of this change on environmental chemistry, but there is no question that certain kinds of lakes are vulnerable to biological impoverishment and effects on forests are well known in both Europe and the United States. In the Amazon, excessive deforestation could disrupt the basic hydrological cycle and trigger an irreversible drying trend. That trend in turn could affect the remaining forest and lead to numerous Amazonian extinctions. The increase in atmospheric carbon from burning of fossil fuels as well as from tropical deforestation has serious implications.

Every year a large number of compounds new to nature are produced in laboratories. Some are sufficiently novel that natural systems have not evolved a capacity to cope with and assimilate them. Among the best known and dramatic are the chlorinated hydrocarbons, most notably DDT. These were well on their way to eliminating some species before the problem was recognized and corrected. Great care is taken to test substances destined for human consumption; similar precautions should be taken for substances destined for environmental consumption.

False hope is sometimes derived from noting the few species which have benefited from conversion of our landscape, such as raccoons which have reached nuisance levels in some eastern cities of the United States such as Washington, D.C. It is easy to let such examples suggest the problem cannot be as bad as portrayed. Yet such species are a tiny minority and their success in modified environments makes them what biologists term "weedy" species, i.e., potential pests.

Not only is there a large and unprecedented number of endangered or potentially endangered species, but this also implies an impending net reduction in global biological diversity. While there have been notable episodes of species extinction in the past, during the history of life on earth, overall there has been a net accumulation of species.[10] Extinction has, more often than not, involved a replacement of one species by another, not necessarily closely related to it. Even if the disappearance of much of the Pleistocene megafauna was human-generated,[11] it really only represented a tiny number of extinctions. Consequently, we stand on the brink of an episode of extinctions which is without precedent in the history of humankind.[12]

II. WHY IS IT IMPORTANT?

Assuming that the biota contains ten million species, they then represent ten million successful sets of solutions to a series of biological problems, any one of which could be immensely valuable to us in a number of ways. *Penicillium* mold has the ability to ward off competitive fungi. The French take advantage of this to use goat's milk to produce a cheese (Roquefort) having distinctive flavor and not susceptible to spoilage by other fungi. This same characteristic is used to combat fungal infections of our bodies or those of our domestic animals. This kind of ability was first observed in a *Penicillium*, which thus provided an impetus for studies of antibiotic activity and resulting medicines. United States superiority in antibiotic medicine was one of many, but nonetheless one of the factors leading to the Allied victory in the Second World War.

Our history is rich in examples of highly useful discoveries based on species previously perceived as worthless. Aspirin, as standard an ingredient of most modern medicine cabinets as any, consists of an organic molecule originally derived from a willow; indeed its chemical name, salicylic acid, is derived from the Latin *Salix* for willow. The recent discovery of a powerful antiviral substance from a sea squirt is another such example.[13]

Some discoveries have an enormous impact and some minor. There is nothing about abundance (or, therefore, rarity) that correlates with usefulness. If anything, there are arguments that rare species may be more useful on the average than abundant ones.[14] Much is made of the potential and present contribution of wild species.[15] A recent analysis of the United States economy shows that 10 percent of GNP is derived directly from wild resources.[16]

Yet as important as some of the contributions of wild species have been, are, or will be to human society, the greatest contribution is likely to be in the form of *knowledge*. The pharmaceutical industry is much more likely to attempt to synthesize a new medically useful compound than it is to try to grow the organism in question in large volume, or to harvest it from the wild. This is even more likely with the advent of biological/genetic engineering.

It might seem at first glance that this engineering potential will provide *less* reason to care about protecting biological diversity; in fact it indicates increased importance. Our knowledge of biological systems is so superficial that there is not a single species for which it can be said with confidence that we know it in its entirety and need not retain it for its potential contribution to biological knowledge. As living organisms we have a vested interest in not limiting the growth of that branch of knowledge known as the science of life. For that reason alone we must be concerned with the survival of each and every species.

The point then is not that the "worth" of an obscure species is that it may someday produce a cure for cancer. The point is that the biota as a whole is continually providing us with new ways to improve our biological lot, and that species that may be unimportant on our current assessment of what may be directly useful may be important tomorrow.[17]

Natural aggregations of species (ecosystems) are more than a large collection of genetic material; they also are involved in ecological processes, often of immediate public service value.[18] A dramatic example is the watershed protection provided by tens of thousands of species in the forests around the Panama Canal. Every time the series of locks is filled and emptied one million gallons of fresh water are used. The canal watershed forests guarantee a steady flow of the fresh water needed for this important link in the world economy.

Many of these processes/services arise from the fundamental fact that living organisms inevitably modify their environment. Indeed the atmospheric composition of this planet is different from all others (high oxygen, for example) because of the presence of life. While it is in one sense paradoxical that life cannot exist without affecting its environment, these modifications have often created conditions favorable to life itself. Life is often also actively involved in maintaining these conditions. For example, the tropical forests of the Amazon are dependent on a minimum amount of rainfall, and yet are (as noted earlier) responsible for generating about half the rainfall in the Amazon basin by hydrological recycling. If the forest were replaced by grassland, rainfall would drop considerably (although not by half since the grasses could return some moisture).[19] The minimum rainfall required for tropical rain forest would then only occur in limited places (e.g., near the Atlantic coast or in places of higher altitude). There, biological factors would not dominate; rather, physical features such as proximity to an oceanic source of moisture or condensation of moisture in rising air masses would generate the necessary rain.

On a global basis the biota is involved in maintenance of the great cycles of water, energy, and the elements. There are already signs that cycles of carbon, nitrogen, and sulfur are being disturbed; disturbance of the biota plays a causal role in most instances, which is augmented in some cases by other factors such as burning of fossil fuels (ironically, fossil biota).

III. BIOTIC IMPOVERISHMENT AND ECONOMIC IMPOVERISHMENT

The foregoing arguments all carry obvious economic implications. Yet the most worrisome aspect of the endangered species problem comes from a disturbing observation that exists for the most part independently of the foregoing arguments: those nations which are unsuccessful in maintaining their basic diversity of plant and animal life are also the ones least successful in protecting decent standards of living for their people. Notable examples in the Western Hemisphere are El Salvador and Haiti, and the latter is of particular interest in comparison with the Dominican Republic on the other half of the island of Hispaniola.

El Salvador has the highest population density of any non-island nation in the Western Hemisphere. Its original forest cover has been reduced to less than one tenth of its original extent, with an attendant although poorly documented decrease in biological diversity. These population/ natural resource-base problems, exacerbated by a tremendous gap between rich and poor, have led to continuing unrest in El Salvador for decades.

Haiti, the second oldest republic in the Western Hemisphere (1804), was once an incredibly valuable colony. Its exports to France were greater than those of any Spanish colony and greater than all combined exports from the American colonies to England. In 1973 only 9 percent of the forest of this largely mountainous country remained.[20] It has the highest population density of any nation in the hemisphere and the density is even greater if calculated on the basis of people per square km of arable (not of excessive slope) land (490 in 1978). Soil erosion from deforestation has reduced the life span of the only significant hydroelectric project, the Peligre dam, from 50 to 30 years.[21]

Today Haiti is the poorest country in the Western Hemisphere. Its gross domestic product per capita is 25 percent that of the Dominican Republic. It has less energy per capita, and a less favorable doctor per population ratio, with attendant higher mortality rates.[22] The Dominican Republic, while no perfect example of environmental management, has lost less of its natural environment and biological diversity. Recognizing the importance of forests for watershed protection, the Dominican Republic put the army in charge of a no-cutting law instituted in synchrony with hydroelectric development.[23]

The comparison is not clear-cut, nor would we expect it to be, whichever two countries might be chosen. For example, the Dominican Republic has twice the land area for roughly the same population. Yet it is sufficient to raise the worrying question about the relation between human welfare and the biological diversity of the landscape.

In general it is easy to understand how there might be such a relationship. By the time environmental problems begin to be widely noticed and diversity begins to decline, there usually is a substantial amount of social and economic momentum which is hard to slow down, let alone halt. Rising population, so often part of the problem, is often coupled with a demographic structure which virtually guarantees further increase. Whatever environmental problems have been noted are therefore almost inevitably going to get worse. This represents a general tendency to overshoot ecologically. By the time the problems are recognized, the drive of society usually has led to increased pressure on marginal lands. This constitutes a feedback system whereby environmental problems themselves tend to create further problems. Ingrained trends in the society are recognized as problems only after they have placed a strain on the biotic environment, but these trends can be expected to continue and worsen for years after they are first noticed.

In any immediate sense it is hard to see how the loss of a single species will affect the day-to-day life of the average citizen who will probably not even know of the loss or that the species ever existed. Yet, when an ecosystem has been altered to the point that a public service benefit such as watershed protection is affected, it is not only quite obvious but, also, biological diversity will have by then suffered a significant reduction. Problems may include loss of clean and reliable water supplies, loss of soil fertility through erosion, and siltation of reservoirs, and they all occur with the impoverishment of the biota. These are measurable in economic terms most simply by the costs of public service deterioration and restoration costs which society would otherwise not have to bear.

It is important to approach this topic by asking the question: Could a nation forfeit a significant portion of its biological diversity yet maintain its living standards? In theory, provided most of the loss was incurred without major degradation of the public services of wildlands, it might be possible. Yet this would require a greater degree of careful environmental planning than is normal or even possible in any practical sense for human societies. Some of the island nations of Southeast Asia (e.g., Bali) may have once approached this. The more usual situation, however, is that by the time problems are recognized, ecological overshoot has already occurred and the situation continues to deteriorate with reversal slow, if possible at all.

Yet, given that sustainable development accompanied by a significant reduction in biological diversity is unlikely to happen, the question still remains, is it possible? One way it might work would be for a highly industrialized nation to depend on reasonably cheap energy supplies and barter products of technology for natural resources. Japan would seem to fit this model. Yet resource limitations would suggest a limit on how many countries could actually take such an approach. This is a problem at the heart of the North-South dialogue between developed and developing nations. How many Japans can there be relative to nations supplying raw materials?

Would it be possible to maintain living standards in the face of significantly diminished biological diversity in an essentially isolated nation? If taken to the extreme of an unrealistically physically uniform landscape, is it possible to design a world that is wall-to-wall wheat fields? The major factor in such an answer is the extent to which ecological "public services" could be maintained by a series of highly simplified human-dominated ecosystems. Yet consideration of public services in such an instance would need to include factors transcending national borders such as maintenance of global cycles of elements, water, and heat. The design and management of such a uniform landscape would be an enormous challenge for science and society. Even if considered a desirable goal, it is unlikely to be achieved in the near future.

As impractical as such a goal may be, biological impoverishment proceeds apace. It does so because, even when considered, endangered species problems are approached individually in financial terms.[24] This is generally attributed to the inability of markets to reflect what are appropriately (in the sense that they are left out of the economic calculus) termed externalities. It is generally either held that externalities can be dealt with in terms of policy, or assumed that if by chance there is some way to calculate values and costs they can be included in the economic analysis. Usually, however, the condition that makes biological resources so valuable, namely, their seemingly infinite but not fully explored potential to support people, becomes a liability when it comes to defending them in financial terms. How can one evaluate something with an economic value yet to be perceived? Without an easily identified dollar-and-cents value, can a resource be worthwhile?

One interesting relationship between biological resources and economics is a tendency toward increased value with increasing rarity. Given a steady demand, a decreasing supply tends to accord higher value to each unit of supply. In certain instances such as in whaling, this had led, together with the heavy investment in the whaling fleet, to making it worthwhile to pursue whales close to biological extinction—but not quite. Economic extinction comes sooner than biological only because the last few individuals are so scattered and hard to find that it becomes inordinately expensive to do so, at least in terms of energy.[25]

This supply/demand relationship means there is an important link between cost of living and dwindling natural resources. Whale products do in fact get more expensive the rarer whales are. In colonial New England, shad was so abundant it was illegal to serve it to servants more than twice a week. Whereas today, with shad rarer and demand greater, it is a highly prized and priced spring delicacy. Similar situations have occurred involving the prices of lobster and salmon.

Admittedly it is difficult to sort out the effect of decreased availability of a particular resource from increased demand. Yet it is certainly the supply/demand ratio that is functioning here. The general thesis should hold true, given the supply/demand model, that dwindling natural resources must be fueling rising living costs and a concomitantly falling standard of living. This is a form of economic effect that is reversible only with the restoration of the populations of the particular economic species.[26] Further, it is an effect that no doubt occurs in subtle forms in growing industrial societies, besides being part of the economic travails of nations like El Salvador and Haiti, where ecological overshoot has already happened.

IV. INCREMENT VS. AGGREGATE

Yet the loss of a single species out of the millions that exist seems of so little consequence. The problem is a classic one in philosophy; increments seem so negligible, yet in aggregate they are highly significant. Accordingly, endangered species with widely recognized economic value are likely to receive some consideration when threatened. But most endangered species are not of immediate economic value. Consequently there will be very few instances when a choice (and many such choices in the United States since the passage of the Endangered Species Act have proven to be only apparent) between an endangered species and a development project will be made in favor of the endangered species. It is very difficult to marshal arguments showing that the particular endangered species will confer more benefits on humanity than the particular development.

Yet, if this reasoning is taken to its logical conclusion, most endangered species will become extinct and the planet will be significantly impoverished biologically, with severe consequences for the welfare of people. The incrementalism problem (where what seems so sensible increment by increment is actually contributing to the development of a major problem) is by no means confined to environmental or endangered species issues. It is a widespread problem in human affairs, and a major reason why there are governments and regulations.

In academic philosophy, it is often argued that the increment vs. aggregate problem does not exist. It is held that each incremental decision creates a new situation within which the next decision is made. But when the increments are in singletons, tens, or even thousands of species out of millions, such effects may be imperceptible, and may seem even more so when many of the effects are delayed or are impossible to measure (e.g., knowledge lost). By the time the accumulated effects of many such incremental decisions are perceived, an overshoot problem is at hand. Thus, even if one accepts the classic understanding of the increment vs. aggregate problem, in conservation, as in other spheres, delayed perception can produce disasters when decisions are made in individual cases with no regard for larger trends.

V. Conservation in Practice

The reality of the imbalance between the pressures on the biota and the meager resources available to rescue the full array of plant and animal species in danger or soon to be in danger must be faced. It has been suggested that it would be sensible to make some thoughtful decisions as to which species to save and which to let go.[27] Such an approach was labeled triage in reference to the World War I medical system of allocating scarce medical supplies and help to those most likely to benefit. Even though the question was originally asked in a rhetorical sense, with the intent to awaken the public to the enormity of the problem of biotic impoverishment, how sensible at first it seems. If there are ten million species, surely it won't hurt to let a few go. Perhaps it won't hurt or hurt much, but since the number is certain to be greater than a few, some rational picking and choosing would seem in order.

Certainly it does make sense to safeguard those which have already clearly been determined as important, such as the wild relatives of our domestic crops. Beyond that it gets difficult and problematical. As noted, the biota is less than half described and we are not yet in a position to state the full worth of even the described species. Furthermore, from a practical point of view, even were omniscience possible, there is no way to separate all the "useful" from the "nonuseful" species; useful species are not found in a single place nor in a single kind of ecosystem. Also, the easiest way to protect a useful species is by protecting its natural habitat, i.e., all the other species that are part of its ecosystem. Further, since it is desirable to protect a valued species in its full variety, it is important to conserve it in all the different places and ecosystem variants in which it occurs. All the foregoing does not even take into account the need to protect ecological processes and public service functions.

In short, triage is both unworkable and misleading in its apparent common sense. Yet what are highly limited conservation agencies to do? It does make some sense to channel resources toward areas that include large numbers of hitherto unprotected species; this and other factors lead to an emphasis on the tropical forests. Beyond that the matter gets into difficult questions as to how much of a particular ecosystem should be protected and, inevitably, how to increase the funding base for conservation while decreasing the human pressures on the natural environment.

There is also a vital need for an international approach to conservation. One major reason is the skewed distribution of financial and technical capacity to carry out such work as compared to actual conservation needs. With conservation in the tropics so much a priority, there are inevitably nations with insufficient manpower or financial resources to address their own conservation problems. Indeed one disturbing aspect of the situation is the question of whether the world in its entirety has enough of the right kinds of scientists to do the job.

Ignoring that problem, the question can be put another way: If each nation took care of its own flora and fauna, would that in fact solve all problems? That would probably mitigate many of them, but it would not necessarily take care of all, in that some problems are inherently international in nature and only careful international coordination can make solutions work. There is no marine turtle, for example, that goes through its entire life cycle in the waters of a single nation.[28] The incredible multigenerational migration of the monarch butterfly from the United States and Canada east of the Rockies to a handful of very tiny spots (each a few hectares at most) in central Mexico is dependent on a complex series of factors in both Mexico and the United States.

The problem of endangered species is common to all nations. Of all incremental problems, the impoverishment of the biota and dwindling biological resources are among the most serious. Indeed, if one considers that declining standards of living contribute to social, economic, and political instability, including increased likelihood of nuclear war, the endangered species problem may well be the most serious. Endangered species serve as heralds of these other, incrementally developing problems.

Our general perception has been that the natural world is there to be exploited or to be a source of amusement. Yet our well-being is far more intertwined with the rest of the biota than many of us would be inclined to believe. I can only conclude that every species, endangered or not, is important, and that the best use of the land in any nation is one that protects its natural biological diversity. Surely there is no reason that the very capacities which have led us to the brink of a major reduction in global biological diversity cannot be redirected toward more sensible use of the land, water, and air, and toward more cognizance of the intertwined destiny of all species.

NOTES

1. John Bartram, of Philadelphia, was founder of the first true botanical garden in America, and the first to inform the Old World about the Venus flytrap. Not surprisingly he was disappointed when, in 1764, William Young, Jr., the 22-year-old son of a German neighbor, was summoned to London and appointed Queen's Botantist with an annual stipend of £300. Both George III and Queen Charlotte (from the German Duchy of Mechlenberg-Strelitz and for whom *Strelitzia regina*, the Bird of Paradise flower, was named) were

enthusiastic amateur botanists. Franklin, along with Bartram one of the nine founding members of the American Philosophical Society, interceded on his behalf and in 1765 Bartram was appointed King's Botanist for North America at the disappointing annual stipend of £50. In gratitude Bartram named this late summer flowering tree, which he had discovered that very year, in Franklin's honor.

2. Eric Eckholm, "Disappearing Species: The Social Challenge," *Worldwatch Paper* 22 (1978); Paul R. and Anne H. Erhlich, *Extinction* (New York: Random House, 1981), p. 306; Thomas E. Lovejoy, "A Projection of Species Extinctions," in *The Global 2000 Report to the President*, prepared by The Council on Environmental Quality, Gerald O. Barney, Study Director (Washington, D.C.: U.S. Government Printing Office, 1980), pp. 327-32; Norman Myers, *The Sinking Ark* (New York: Pergamon Press, 1979), p. 307; Norman Myers, "Conversion of Moist Tropical Forests," A Report prepared for the Committee on Research Priorities in Tropical Biology of the National Research Council (Washington, D.C.: National Academy of Sciences, 1980); Peter H. Raven (Chairman), *Research Priorities in Tropical Biology* (Washington, D.C.: National Academy of Sciences, 1980); Edward O. Wilson, "Resolution for the Eighties," *Harvard Magazine* (1980): 20-25, 70.
3. Michael Harwood, "Math of Extinction," *Audubon Magazine* 84 (1982): 18-21; Herman Kahn and Julian L. Simon, eds., *The Resourceful Earth* (Oxford: Basil Blackwell, 1984), p. 585.
4. Myers, "Conversion of Moist Tropical Forests."
5. Kahn and Simon, *The Resourceful Earth*, p. 585.
6. Peter H. Raven, "Trends, Priorities, and Needs in Systematic and Evolutionary Biology," *Systematic Zoology* 23 (1974): 416-39; Terry L. Erwin, "Tropical Forests: Their Richness in Coleoptera and Other Arthropod Species," *Coleopterists Bulletin* 36 (1982): 74-75.
7. Lovejoy, "A Projection."
8. Thomas E. Lovejoy and David C. Oren, "The Minimum Critical Size of Ecosystems," in *Forest Island Dynamics in Man-Dominated Landscapes*, edited by Robert L. Burgess and David M. Sharpe (New York: Springer-Verlag, 1981), pp. 7-12; Thomas E. Lovejoy, Judy M. Rankin, Richard O. Bierregaard, Jr., Keith S. Brown, Louise H. Emmons, and Martha E. Van der Voort, "Ecosystem Decay of Amazon Forest Remnants," in *Extinctions*, edited by Matthew H. Nitecki (Chicago: University of Chicago Press, 1984), pp. 295-325; Daniel Simberloff, "Big Advantages of Small Refuges," *Natural History* 91 (1982): 6-14; John Terborgh, "Preservation of Natural Diversity: The Problem of Extinction Prone Species," *BioScience* 24 (1974): 715-22; John Terborgh, "Faunal Equilibria and the Design of Wildlife Preserves," in *Tropical Ecological Systems: Trends in Terrestrial and Aquatic Research*, edited by F. B. Golley and E. Medina (New York: Springer-Verlag, 1975), pp. 369-80; Robert L. Burgess and David M. Sharpe, eds., *Forest Island Dynamics in Man-Dominated Landscapes* (New York: Springer-Verlag, 1981), p. 310; Jared M. Diamond and Robert M. May, "Island Biogeography

and Design of Natural Reserves," in *Theoretical Ecology: Principles and Applications* (Oxford: Blackwell Scientific Publications, 1976), pp. 163-86; Edward O. Wilson and Edwin O. Willis, "Applied Biogeography," in *Ecology and Evolution of Communities*, edited by Martin L. Cody and Jared M. Diamond (Cambridge, Mass.: Harvard University Press, 1975), pp. 522-34.

9. R. Herbert Bormann, "The Vulnerable Landscape," *Yale Alumni Magazine* 45 (1982): 10-17; T. C. Hutchinson and M. Havas, eds., *Effects of Acid Precipitation on Terrestrial Ecosystems* (New York: Plenum Press, 1980); E. B. Cowling, "Acid Rain: An Emerging Ecological Issue," in *Issues and Current Studies* (Washington, D.C.: National Research Council, 1980), pp. 78-87.

10. Luis W. Alvarez, Walter Alvarez, Frank Asaro, and Helen V. Michel, "Extraterrestrial Cause for the Cretaceous-Tertiary Extinction," *Science* 208 (1980): 1095-1108.

11. That the extinctions of the Pleistocene megafauna were caused by early man is regarded by some as only a matter of circumstantial evidence. Certainly there are some cases, such as the moas of New Zealand, where that evidence is pretty strong. For a recent discussion see Paul S. Martin, "Catastrophic Extinctions and Late Pleistocene Blitzkrieg: Two Radiocarbon Tests," in Nitecki, *Extinctions*, pp. 153-89.

12. A recent hypothesis suggests that there have been fairly regular episodes of mass extinctions with the periodicity of 26 million years (David M. Raup and J. John Sepkoski, "Periodicity of Extinctions in the Geologic Past," *Proceedings of the National Academy of Sciences* 81 (1984): 801-805. This claim has, however, been questioned on a number of grounds. See the following articles which all appeared in *Nature* 308 (April 19, 1984): J. Maddox, "Extinctions by Catastrophe?" p. 685; M. R. Rampino and R. B. Strothers, "Terrestrial Mass Extinctions, Cometary Impacts and the Sun's Motion Perpendicular to the Galactic Plane," pp. 709-12; R. D. Schwartz and P. B. James, "Periodic Mass Extinctions and the Sun's Oscillation about the Galactic Plane," pp. 712-13; D. P. Whitmire and A. A. Jackson IV, "Are Periodic Mass Extinctions Driven by a Distant Solar Companion?" pp. 713-15; M. Davis, P. Hut, and R. A. Muller, "Extinction of Species by Periodic Comet Showers," pp. 715-17; W. Alvarez and R. A. Muller, "Evidence from Crater Ages for Periodic Impacts on the Earth," pp. 718-20. Even if periodic mass extinctions prove to be real, the episode that human societies could cause is premature for the cycle and not in our best interests.

13. Kenneth L. Rinehart, Jr., James B. Gloer, Robert G. Hughes, Jr., Harold E. Renis, J. Patrick McGovern, Everett B. Swynenberg, Dale A. Stringfellow, Sandra L. Kuentzel, Li H. Li, "Didemnins: Antiviral and Antitumor Depsipeptides from a Caribbean Tunicate," *Science* 212 (1981): 933-35.

14. See Bryan G. Norton, and Geerat J. Vermeij, both this volume.

15. Norman Myers, *A Wealth of Wild Species* (Boulder, Colorado: Westview Press, 1984), p. 274.

16. Robert and Christine Prescott-Allen, *The First Resource* (New Haven, Conn.: Yale University Press, 1986).
17. See Bryan Norton, this volume.
18. Ehrlich and Ehrlich, *Extinction*, chap. 5.
19. Eneas Salati and Peter B. Vose, "Amazon Basin: A System in Equilibrium," *Science* 225 (1984): 129-38.
20. "Haiti: A Study in Environmental Destruction," *Conservation Foundation Newsletter* (November 1977).
21. "Draft Environmental Report on Haiti," prepared by Science and Technology Division, Library of Congress, for U.S. Agency for International Development (Washington, D.C.: January 1979).
22. James W. Wilkie and Stephen Haber, eds., "Statistical Abstract of Latin America," Vol. 21 (Los Angeles: Latin American Center, University of California, Los Angeles).
23. Ian Bell, *The Dominican Republic* (Boulder, Colorado: Westview Press, 1981).
24. Norton, this volume.
25. Colin W. Clark, "The Economics of Overexploitation," *Science* 181 (1973): 630-34; Colin W. Clark, *Mathematical Bioeconomics: The Optimal Management of Renewable Resources* (New York: Wiley, 1976), p. 352.
26. Lester R. Brown, *Building A Sustainable Society* (New York: Norton, 1981), pp. 288-91.
27. Thomas E. Lovejoy, "We Must Decide Which Species Will Go Forever," *Smithsonian Magazine* 7, no. 4 (1976): 52-58.
28. Archie Carr, Jr., pers. comm.

2

The Biology of Human-Caused Extinction

GEERAT J. VERMEIJ

I. INTRODUCTION

Man is responsible for the extinction of many species. Moreover, his role as biological exterminator is unlikely to diminish in the foreseeable future. Three critical questions must be answered in order to understand the implications of extinction for human welfare. These are: (1) Which kinds of species are susceptible to extinction, and which are not? (2) How will the extinction of species affect the communities in which these species lived and alter the evolutionary environment of surviving life forms? (3) Can new species evolve on a man-dominated planet to replace the species which have disappeared and, if so, what will these new species look like? The answers to these questions will determine our views about the values of species in general and of endangered species and communities in particular. Without a profound scientific understanding of extinction, rational policies about the conservation of diversity cannot exist. It is important to take as dispassionate a view of extinction as possible, and not to fuel the sensationalism that has pervaded much of the popular literature on this topic.

In this chapter I shall try to summarize what is known and what is not known about the causes, selectivity, and consequences of extinction. I shall argue that extinctions in the geological past provide insights and predictions about human-caused extinctions. Theoretical considerations and some data will be adduced to suggest that species for which selection from biological sources is important are more susceptible to extinction than are other kinds of species. The disappearance of such species will have important effects on the evolutionary potential of surviving species. These indirect consequences of extinction, I shall argue, may be as profound as are the extinctions themselves. Finally, I shall try to show that, although extinction and the evolution of new species are favored by the same conditions of habitat fragmentation, extinction is the more likely outcome given the habitat destruction being wrought by man. A future man-dominated world will be biologically more homogeneous and less rich in forms adapted to competition and predation than is the world today.

II. Causes of Extinction

No attempt to reduce the death toll of species can succeed until the causes of extinction are better understood. Predictions about which species are at greatest risk depend on which causes of extinction are more important.

It is useful from the outset to distinguish between causes for the decline of thriving populations and causes for the demise of already severely depleted populations. Very small local populations can succumb to events that would not result in the extinction of larger or more dispersed populations. Emphasis on the somewhat haphazard reasons for the extinction of decimated populations would therefore divert attention from the more important and tractable question of how large populations become depleted.

In the broadest sense, extinction occurs when populations cannot persist in the face of environmental change.[1] This change can be physical (unusual weather, pollution, soil erosion, habitat destruction) or biological (the addition or elimination of competitors, predators, parasites, prey, or symbionts). Extinction is only one of three reactions a population can have to a change in conditions. Alternatively, a population may migrate to areas where the environment is not changing, or it may adapt to the changing conditions. How species react to change and which characteristics influence the "choice" are central questions in evolutionary biology to which I shall return briefly later in this essay.

Perhaps the best studied form of physical degradation is the fragmentation of once continuous habitat into island-like refuges. When Barro Colorado became an island in Gatun Lake with the completion of the Panama Canal in 1914, a reduction in bird diversity began which has, as of 1981, resulted in the local extinction of at least 50 forest birds (about 20 percent of the original 237 species).[2] Destruction of forests in southern Brazil has resulted in the disappearance of many birds.[3]

Another instructive example of large-scale extinction due to physical degradation is the decimation of the freshwater biota of rivers in the central and southeastern United States. Deforestation, pollution, and especially the building of dams have altered freshwater habitats and largely

29

removed swift-water shallows in the larger rivers. At least eight species of the unionid bivalve genus *Dysnomia* and all five species of an endemic subfamily (Neoplanorbinae) of ancylid pulmonate limpets have become extinct, as have many local races of gastropods.[4]

More than 30 of 103 species of bivalve in the Ohio River drainage are considered to be endangered in that they persist only in a few relict populations; this number has probably increased since Stansbery's status report.[5] The loss of this freshwater fauna is particularly unfortunate because many of the species showed remarkable adaptations to predators. No other fauna of river molluscs was so large or so highly adapted to biological agents of natural selection as was (and to some extent is) the North American freshwater fauna.[6] Today, many of the species are being replaced by weedy introduced species such as the European snail *Bithynia tentaculata* and the Asiatic clam *Corbicula manillensis*.

In Israel, the draining of marshes and destruction of Mediterranean scrub and shrub habitats caused 26 plants to become locally extinct in the 30 years prior to 1976.[7] This number represents about 1 percent of the Israeli vascular flora and a much higher proportion of plants in the affected habitats. This is undoubtedly not an unusually high figure and more documentation for other floras is clearly desirable.

Although man's activities are causing habitat fragmentation on an unprecedented scale, fluctuations in habitat size are not new. Changes in sea level as glaciers advanced and subsided during the Pleistocene epoch caused large changes in the size of islands and in the distances between them. Episodes of habitat coalescence and fragmentation were accompanied by climatic shifts. In the West Indies, for example, arid conditions were evidently far more widespread than they are today, and many vertebrates which were adapted to these conditions have become extinct or greatly reduced.[8] The rate of extinction of large mammals was high throughout the Pleistocene among North American mammals because the advance of glaciers periodically made habitable areas for animals with large individual ranges much more limited.[9]

Even in the sea, where human-caused extinctions are fortunately still rare, cooling and perhaps other factors caused a 20 percent reduction of the molluscan fauna of Florida and the Atlantic coast north to Maryland during the Early Pleistocene, 1.1 to 1.8 million years ago.[10] At the same time, extinction in the North Sea Basin gastropod fauna, based on my calculations from the Early Pleistocene of the Netherlands, was 50 percent.[11]

Extinction from biological causes may come about in either of two ways. In the first, predators or competitors evolve through time to improve their capabilities, so that species which were already vulnerable as prey or subordinate competitors become even more susceptible. Man's acquisition of weapons is an extreme example of predatory improvement against which most already rare large mammals were unable to respond evolutionarily. Consider what happened to mammals in the Americas after man arrived during the Late Pleistocene. Marshall, in his incisive analysis of the history of South American mammals, estimates that 45 of 120 genera (38 percent) of South American land mammals became extinct about 15,000 years ago, some time after humans first came to that continent. This proportion corresponds to an extinction rate of 150 genera per million years, a value four times higher than in any other Late Tertiary or Pleistocene time interval.[12] In North America at about the same time, 32 of 114 genera of land mammals (28 percent) became extinct; this proportion corresponds to an extinction rate of about 46 genera per million years, a value 2.5 times as high as in earlier Late Cenozoic time intervals.[13] The vast majority (96 percent) of these extinctions affected large-bodied mammals.

Many other extinctions are also attributable to human hunting. The six species of bird which have become extinct in North America since 1600 were all decimated by hunters.[14] The same is true of the extinct large marine mammals such as Steller's sea cow (*Hydrodamalis gigas*). Extinction of giant tortoises on islands (11 species and subspecies), large West Indian lizards (6 species) and snakes (4 species), and many insular flightless birds is attributable in part to human exploitation.[15] Ehrlich and Ehrlich discuss instances of people removing all individuals of endangered cactus species from islands off the coast of Baja California.[16]

Why could these species not respond evolutionarily to improvements in man's weaponry? Effective antipredatory defenses are maintained only when a substantial portion of a prey population is exposed to and survives predatory attack.[17] I suspect that man's predatory efficiency rose dramatically each time new weapons such as arrows, spears, and firearms were invented, and as transportation improved. Not only did fewer animals which were attacked survive to reproduce, but most members of the population were eventually attacked and killed by humans. As a result, there were few places where the population could perpetuate itself in the absence of man, and selection for the maintenance and improvement of defense became weaker. This hypothesis depends on the observation that human predatory efficiency increased with the development of more sophisticated weapons. Accordingly, it will be of great interest to gather data on predatory efficiency for various types of weapon. I do not know of such data in the anthropological literature.

A second, and from the human perspective a more indirect, biological cause of extinction is invasion by foreign species. In terrestrial commu-

nities, the most devastating invaders are species which man has introduced either deliberately or by accident. The introduction of goats, rats,
mongooses, and mosquitoes has been particularly destructive to the biota
of ocean islands. Warner believes that the near destruction of the Hawaiian fauna of forest birds (85 percent of the endemic species have
become extinct or greatly reduced since 1850) resulted from the spread
of bird pox and avian malaria by the mosquito *Culex quinquefasciata*,
which was introduced to the islands in about 1826.[18]

Atkinson considers the roof rat (*Rattus rattus*) to be the chief culprit.[19]
After its introduction to Oahu at some time between 1850 and 1870,
the rat increased in population size as the number of forest birds declined
sharply. The principal period of decline was later on the other islands
and may correspond to population surges of the rat. During the 20 years
after the first appearance of the roof rat on Lord Howe Island in 1918,
five bird species (33 percent of the original fauna) became extinct.[20]

By reducing vegetation cover and by eating the leaves of plants, goats
which were released by European explorers have undoubtedly been responsible for the extinction of many endemic island plants and may have
contributed indirectly to the extinction of some animals.[21] On Ascension
Island, three or four of the ten vascular plant endemics have become
extinct either because of grazing by goats and sheep or because of competition with African grass (*Melinis* spp.) or any of the 300 other introduced plant species.[22]

Several authors have seen a connection between the human exploitation
of native species and the subsequent establishment of introduced species
which in turn depress the stocks of native species still further. Christie,
for example, contends that overfishing of native whitefishes (*Coregonus*
spp.) and other species in the American Great Lakes contributed to the
successful establishment of such introduced fishes as the smelt *Osmerus
mordax*, the lamprey *Petromyzon marinus*, and the alewife *Alosa pseu-
doharengus*, which further reduced native fish populations.[23] In New
Zealand, native birds have disappeared from the mainland forest, which
is relatively more open and more disturbed by humans and introduced
animals like goats than are the forests of some offshore islands. Diamond
and Veitsch, who observed this pattern, suggest that the 34 introduced
bird species with established populations are far more successful in pen-
etrating the mainland forests than the denser forests of the offshore
islands.[24] The replacement of extinct species by introduced ones may thus
be indirect and mediated by human disturbance and exploitation. The
34 introduced bird species do not precisely replace the species (31 percent
of the pre-Maori fauna) which have become extinct in New Zealand.[25]

As Simberloff has pointed out, the introduction of foreign species rarely results in the total eradication of species on larger continents or in large marine biotas.[26] Many species which have been introduced by man from Europe to the Americas and Australia, including the house sparrow (*Passer domesticus*), starling (*Sturnus vulgaris*), brown and roof rats (*Rattus norvegicus* and *R. rattus* respectively), and house mice (*Mus musculus* and *M. mus*), are intimately associated with the habitations of man, and therefore do not, under most circumstances, interact with native species in habitats away from human settlement. Other introduced species, such as rabbits (*Oryctolagus cuniculus*), annual weedy flowering plants, the tropical American tangantangan tree (*Leucaena leucocephala*, introduced to Pacific islands for control of soil erosion), and many others, are characteristic of disturbed environments in which these species might be outcompeted or consumed by predators if the habitats were left to themselves.[27] Most marine animals introduced by man, such as the American slipper limpet (*Crepidula fornicata*) in Europe, the New Zealand barnacle *Elminius modestus* in Europe, and the numerous species from Europe, eastern North America, and Japan now in San Francisco Bay, are organisms found in sheltered bays and estuaries.[28] Where their impact has been studied, as for the Japanese snail *Batillaria zonalis* in San Francisco Bay,[29] the European periwinkle *Littorina littorea* and green crab *Carcinus maenas* on the north Atlantic coast of America,[30] and the southwest Pacific snail *Trochus niloticus* in Guam,[31] some native species may have suffered population declines because of competitive inferiority or vulnerability as prey, but none has disappeared. The 50 species of fish which have been introduced successfully to streams and rivers in California may have reduced populations of native species, but only one species (the Sacramento perch, *Archoplites interruptus*) has become locally extinct as a consequence of competition with an introduced species (the blue-gill sunfish, *Lepomis macrochirus*).[32] The Sacramento perch survives as an introduced species in several streams where blue-gills have remained rare. Moyle points to three examples of genetic swamping of local subspecies of fish as a result of the transplantation of closely related subspecies from adjacent drainages.[33] Finally, the introduction of the predaceous fish *Cichla ocellaris* for the pleasure of sportfishermen in Gatun Lake, Panama, has had profound ecological effects including a reduction in populations of native fishes, water birds, and planktonic Crustacea, and a possible increase in malaria-carrying mosquitoes, but local extinction has been rare.[34]

Perhaps the most ambitious biogeographical experiment yet performed by man is the construction of the Suez Canal, which opened for navigation in 1869. Since that time, at least 36 fishes, 68 molluscs, 20 decapod crustaceans, and many other organisms have migrated through the canal, overwhelmingly from the Red Sea to the Mediterranean.[35] The establishment in the Mediterranean of predaceous fishes from the Red Sea has caused population shifts of native Mediterranean prey and competitors,[36] but I know of no confirmed example of extinction of any native Eastern Mediterranean species. Achituv's inability to find the native sea-star *Asterina gibbosa* on the Mediterranean coast of Israel after migration of *A. wega* from the Red Sea in 1955 may point to one case of local extinction, but the evidence is not yet compelling.[37]

An even more elaborate scheme to mix previously isolated biotas is being contemplated in some quarters. The construction of a sea-level saltwater canal connecting the tropical Western Atlantic with the tropical Eastern Pacific in Central America would reunite two highly diverse biotas which have been effectively isolated and diverged from one another for 3.1 to 3.5 million years. Present evidence suggests that the Eastern Pacific rocky-shore fauna has a higher incidence and greater development of antipredatory adaptations than does the corresponding Western Atlantic fauna.[38] Moreover, many species which are specialized to eat stony corals are found in the Eastern Pacific but not in the Atlantic. Transport of these animals, including the crown-of-thorns sea-star *Acanthaster planci*, could pose serious threats to Caribbean reefs, whose diversity and development are currently much greater than those in the Eastern Pacific.[39]

Exchange of species between previously isolated biotas has a long history. The best studied case is the migration of mammals between North and South America after uplift of the Panama land bridge a little more than three million years ago.[40] Although the establishment of North American mammals in South America is broadly coincident with the extinction of many South American natives, particularly among herbivorous species belonging to endemic orders, climatic changes also appear to have played a role in this replacement, and the precise biological reasons for the demise of the native forms (predation, competition, or disease) remain unknown. Other cases of migration include the late Cenozoic colonization of the North Atlantic by species from the North Pacific via the Arctic,[41] the relatively recent spread of Indo-West-Pacific reef-associated species to the Eastern Pacific,[42] and the migration of terrestrial organisms between Asia and Alaska.[43] At least in the marine examples, extinction may have been a rare consequence of these migrations.

The scant data lead me to the conclusion that habitat fragmentation and destruction and human hunting are the most important causes of current extinctions. The invasion of foreign species appears to have had destructive effects chiefly on islands and in other situations where species in the recipient communities have small populations or much less well-developed defenses against competitors and predators than do the introduced forms. Large-scale mixing of previously isolated biotas might prove calamitous, especially if the two biotas differ in the development of biological defenses.

III. SELECTIVITY OF EXTINCTION

Vanished species are unlikely to be a random subset of the biota in which they lived. In their review of fragmentation of habitats on land, Terborgh and Winter concluded that the species most susceptible to extinction are constitutively rare, that is, have very low population densities and large individual ranges.[44] Large mammals require very large home ranges and probably cannot maintain populations even in relatively large refuges. Several large predators, including the puma, jaguar, and harpy eagle, have disappeared from Barro Colorado Island, and have become very rare in adjacent mainland forests in central Panama.[45] In his analysis of the extinction of birds on Barro Colorado, Karr found that ground-dwelling birds suffered a much higher rate of extinction than did canopy species, and that species typical of foothill habitats were more vulnerable than those of dry and wet lowland forest.[46] He attributed these differences to two principal factors. First, ground-dwelling and foothill birds depend on moisture, which is available in limited quantities in the dry season on the mainland but is absent in some very dry years on Barro Colorado. Birds on the island in search of food or moisture might face starvation in the dry season because they cannot cross the narrow water barrier to the mainland. This low dispersibility, the second factor influencing extinction, acts in concert with climatic stress to prevent the reintroduction of members of the mainland population to the island. Canopy birds have greater dispersibility and are less susceptible to local extinction.

These studies indicate that a given climatic stress or episode of fragmentation does not affect all species equally. In general, large species should be more susceptible to extinction than small ones because their smaller populations are less capable of rapid recovery. Ectothermic ("cold-blooded") animals, with low metabolic rates and long periods of inactivity, should be less susceptible to habitat fragmentation and environmental stress than are endotherms ("warm-blooded" animals), whose high metabolic rates and continuous activity make for intolerance to temporary adversity.[47] This may be the reason why no ectothermic lizard species is known to have become extinct on Barro Colorado,[48] whereas the bird fauna was reduced by about 20 percent. Finally, I expect insect-pollinated flowering plants to be more prone to extinction than wind-pollinated forms. Pollination by insects permits plants to persist at low densities,[49] so that a crisis could easily wipe out a local population. It must be added that the characteristics which increase vulnerability to extinction also promote speciation.[50] This point will be discussed in detail below.

The greater dispersibility, larger population sizes, and presence of more effective refuges among marine organisms as compared to those on land may explain why human-caused extinction has thus far been comparatively rare in the sea, except for large birds and mammals. I agree with Abbott and Keen that few, if any, marine shelled molluscs have become extinct in the last 300 years in spite of the ruthless overexploitation of many species by shell collectors and by people gathering molluscs for food.[51] Some local forms of species with a bottom-dwelling dispersal phase are doubtless being threatened, and E. J. Petuch suggests to me that one or more forms loosely classified as belonging to the locally variable species *Melongena corona* may already be extinct in Florida.

Another way of expressing the idea that marine invertebrates are relatively immune from extinction is that the scale of disturbance which is sufficient to bring about the demise of local populations is larger than that needed to extinguish land vertebrates. A good example of the resistance of small marine invertebrates to total extinction is provided by the large-scale destruction of eelgrass (*Zostera marina*) beds, which began in 1932-33. In certain Danish fjords, the death of eelgrass led to changes in sediment characteristics and shore topography, an increase in the abundance of mussels and barnacles, and the decimation of species which are found chiefly or solely on the leaves of the plant.[52] According to Rasmussen's careful historical investigation, only two species (the anemone *Sagartiogeton viduata* and the sea urchin *Psammechinus miliaris*) became locally extinct in the Isefjord and its arms after the demise of the eelgrass.[53] Even though eelgrass destruction was widespread and contemporaneous throughout the North Atlantic, no species appears to have died out completely. Incidentally, Rasmussen believes high summer temperatures to be the primary cause for the "wasting disease" which culminated in the death of eelgrass. Since the 1930s, eelgrass has again increased in abundance, though it is still absent from many previously occupied areas.

Are there other features of species which increase the likelihood of extinction? I suggested that "biotically competent" species, which have evolved in response to intense selection from biological agents (predators and competitors), are on the average more prone to extinction and to evolutionary change than are stress-tolerant species (those living in conditions where individual activity is severely constrained).[54] In order to understand this hypothesis, it is necessary first to outline the dichotomy between biotically competent and stress-tolerant species. Biotically competent species are most apt to live in areas where there is sufficient light and a sufficient supply of inorganic raw materials to sustain organisms which are capable of synthesizing organic substances and tissues from inorganic components. These areas are thus the sites of primary production which sustains all animal life. In the many environments where life depends exclusively on the importation of organic nutrients, there is inevitably a smaller resource base, because some of the biomass that is produced in the source area has already been removed by animal consumers and decomposers. Animals living in habitats away from sites of primary production must maintain low metabolic rates, so that they do not deplete the available nutrients. Examples of habitats in which animals depend exclusively on the importation of nutrients include the deep sea, caves, and deep levels in soils and sediments. Limited primary production is possible at high latitudes and altitudes, in deserts, in perpetually moist understories of forests, on the upper reaches of the seashore, and on nutrient-depleted soils. Low metabolic rates in environments such as these have important consequences for the extent to which adaptation to predation and competition can occur. Animals with a low metabolic rate are limited in growth rate, muscle power, escape speed, and endurance, and those living in cold habitats are subject to the additional constraint of a low rate of calcification, a process which is important to the production of shells and other forms of antipredatory armor. Moreover, the reduced activity which usually accompanies low metabolic rates means that individuals have relatively few encounters with potential enemies, so that the number of times when selection in favor of competitive superiority or antipredatory characteristics operates is small. In short, individuals are adaptationally less constrained and usually have a greater biotic competence (greater expression of competitive and antipredatory features) in areas where primary productivity is high than in places where primary production is limited or impossible. The evolutionarily important consequence of this difference is that habitats and regions of reduced primary productivity have, through the course of geological time, served as refuges for species whose ancestors were unable to cope with the increasingly rigorous biological environment of competitors and predators in environments of high primary productivity, where adaptation to biological agencies is less constrained.[55]

I can think of at least two reasons why species in areas of high primary production should be especially prone to extinction. The first is that these areas are on land and in shallow water, where the effects of large-scale catastrophes and climatic alterations are apt to be most profound. Secondly, the biological component of selection is high for biotically competent species, so that conditions of life could change even if climates remained constant and catastrophes did not occur. Such changes are probably difficult to overcome by adaptation. The most common consequences of biological and physical change for biotically competent species are therefore likely to be extinction and distributional restriction to refuges.

Another category of organisms which would seem to be especially vulnerable to extinction consists of species in which the rate of recovery after a precipitous population decline is low. If populations remain small for long periods following a crash, they could be wiped out completely during the next crisis. In fact, a crisis might not have to be very unusual for it to effect extinction of a small or local populational remnant. High rates of population increase are permitted in small-bodied species, warm-blooded animals, and species with a high individual growth rate. Large body size, the production of heavy armor, and other characteristics which imply a small equilibrial population size and low individual growth rate are associated with a low potential for rapid recovery of decimated populations. The importance of rapid recovery to the persistence of populations has been stressed by Karr in connection with the disappearance of birds from Barro Colorado Island.[56] Note that large body size and heavy armor are associated with biotic competence as well as with a low intrinsic rate of population recovery. The effects of slow population growth should not, of course, be as critical for species in refuges, where fluctuations in the number of individuals are probably small.

Population growth rate is affected not only by the properties of individual organisms, but also by the productivity of the environment. High population growth rates can be achieved in marine waters which are fertilized by upwelling (a process which permits nutrient-rich waters from great depths to be brought to the surface) and by nutrient-laden runoff from the land. These nutrient-enriched environments may serve as important refuges for biotically competent species which elsewhere might have become extinct.

These hypotheses about the role of biotic competence, population dynamics, metabolic rates, and refuges in extinction are speculative, and have not been thoroughly examined empirically either with regard to the major extinction events of the geological past or with regard to human-caused extinctions. My attempts below to identify patterns should therefore be viewed with caution and skepticism.

Analysis of Early Pleistocene gastropods of the southern North Sea reveals that large species were much more likely to become extinct than were small species. For each species listed by the authors of the catalog of Early Pleistocene species from deposits in Zeeland, the Netherlands, I recorded the largest dimensions given, and I noted whether the species has become globally extinct or whether it has survived to the present day.[57] Of the 66 species less than 20mm in length, 24 (36 percent) became extinct, whereas among the 57 species longer than 20mm, 38 (67 percent have died out.

No selective extinction is detectable among the Early Pleistocene gastropods described by Ladd from the Mariana Limestone of Guam and the Tanapag Limestone of neighboring Saipan in the Mariana Islands of the Western Pacific.[58] Of the 62 species recorded by Ladd, 14 are now locally extinct in the Marianas and 3 have died out completely. I divided the species into two categories, those whose shells have apertural defenses against predators (narrow aperture, teeth on outer lip, thickened outer lip) and those whose shells lack these defenses. The proportion of extinct species in the first category (6 out of 24 species, 25 percent) is very similar to that in the second category (11 out of 38 species, 29 percent). The extinct species are also similar in size to those which still survive in the Marianas today.

In a study of molluscan extinction after the Pliocene epoch in marine tropical America, we found that the presence of antipredatory features was associated with a high probability of extinction of rocky-bottom gastropods which occupied the tropical Western Atlantic Ocean, whereas co-occurring gastropods which lacked antipredatory specializations of the aperture suffered much less extinction.[59] This difference was unrelated to the important effect of a small geographical range, which also was associated with a high probability of extinction. The fauna of the nutrient-rich tropical Eastern Pacific was much less impoverished than that of the Western Atlantic, where there was evidence of nutrient depletion after the Pliocene. In fact, the Eastern Pacific served as an important refuge for gastropods and other molluscs which were once also found in the Western Atlantic.

Too little space is available to document in detail the suggestion that environments in which primary productivity is low or absent are important refuges for many species. The evidence that this is so is, however, strong. Many groups which during the Palaeozoic and Early Mesozoic eras had representatives in shallow marine waters are today found only in the deep sea and in dark marine caves.[60]

These examples from the geological record generally support the idea that biotically competent species are more susceptible to extinction than are other types, except perhaps when the extent of faunal impoverishment is small, as it seems to have been in the tropical Western Pacific during the Pleistocene. The vulnerability of large-bodied species to extinction on land is also consistent with the view that biotically competent species and those with a low rate of population growth are at greater risk than are species with a higher intrinsic rate of recovery.[61] I must emphasize again, however, that this topic requires a great deal more empirical research before the various hypotheses can be said to be well supported.

IV. CONSEQUENCES OF EXTINCTION

The disappearance of a species represents more than merely the loss of a biological entity; it also heralds a change in the selectional environment of surviving species. This change can affect many species if the extinct species interacted with many of its neighbors. Surprisingly, this topic has received little attention from biologists.

A good example of the potentially profound effects which the extinction of a single species can have is the near extinction of the sea otter (*Enhydra lutris*). This mammal was on the brink of total extinction until the mid 1960s, when populations in California and Alaska began a resurgence. Areas in which sea otters feed are dominated by large kelp forests and by relatively low densities of sea urchins, abalones, and other animals which graze or browse kelp and other algae. Moreover, the grazers are confined to cracks and crevices where the sea otter, which feeds heavily on the grazers, cannot find them easily. In the absence of otters, grazers become more abundant, can survive on exposed surfaces, and devour large amounts of kelp.[62] It is likely that kelp forests were more localized when the otter was unexploited by man than they are today, particularly when the Steller's sea cow (*Hydrodamalis gigas*), a kelp grazer which became extinct around 1750, was still common in the North Pacific.[63] The disappearance of the otter thus relaxed antipredatory selection in the invertebrate grazers and changed the character, if not the intensity, of selection for competitive ability and grazer-resistant chemistry in the kelps. Overfishing of lobsters (*Homarus americanus*) in the northwestern Atlantic and of spiny lobsters (*Panulirus interruptus*) in southern California has had similar effects on the distribution and selectional conditions of invertebrates and algae.[64]

When one species upon which another is obligately dependent becomes extinct, the dependent species experiences a change in selection so profound that it may also succumb. Two possible examples, both involving plants whose seeds depend on passage through the digestive tracts of extinct animals for germination, will illustrate this point. Temple has suggested that the very hard nut-like seeds of the sapotaceous tree *Calvaria major* of Mauritius require abrasion of the outer coat in order to germinate.[65] None of the native animals of Mauritius, where the tree is now limited to a few aging individuals, is capable of softening the seed to the point where germination can take place. He suggests that germination depended on the dodo (*Raphus cucullatus*), which became extinct in the late seventeenth century, and which had in its digestive tract stones in the powerful gizzard by means of which seeds could be weakened. A similar suggestion has been made for at least 30 trees and shrubs of dry deciduous forests of Costa Rica.[66] The hard nut-like seeds of these plants today accumulate in rotting fruits around the parent plants, where they are usually killed by beetles. Janzen and Martin speculate that gomphotheres, groundsloths, and other large herbivores, which became extinct in the Late Pleistocene, dispersed the fruits and provided trituration, or grinding up, of the seeds in the gut so that germination could take place.[67] In the absence of the herbivores, the plants have become progressively more restricted ecologically and might well have faced extinction were it not for the introduction of cattle and horses, which evidently fulfill the role of the ancient herbivores.

The Ehrlichs have asserted that the extinction of individual species would upset geochemical cycles and other processes at the community or ecosystem level.[68] As species disappear from a community, they argue, regulation of community-wide processes becomes less precise and might, at some critical point, break down. Recent support for this assertion comes from a detailed geochemical analysis of sediments laid down during and immediately after the Cretaceous-Tertiary boundary event, when the impact of a comet or asteroid may have shut out the sun for a substantial period and brought on the mass reduction and eventual extinction of many phytoplankton species in the surface waters of the ocean.[69] With the disappearance of these calcareous organisms, many animals dependent on phytoplankton for food also became extinct, and the regime of sedimentation, which was dominated in the Late Cretaceous by the deposition of biologically produced limestone, changed to a clay regime. The rediversification of phytoplankters and the organisms dependent on them was a slow process. Pelagic communities did not return to Cretaceous norms of sedimentation and productivity until at least 350,000 years after the biotic crisis.

It may seem inconceivable that species of phytoplankton and zooplankton, which have billions of individuals spread over vast areas of the ocean, could vanish and cause such long-term alterations in geochemical processes. The events of the Cretaceous-Tertiary boundary and others like them during the course of earth history serve as potent reminders that, although the regulation of critical nutrients is for the most part under the control of microorganisms which are usually resistant to total extinction,[70] upheaval of the biosphere can happen again. The effects of such upheavals may be felt for hundreds of thousands of years.

V. The Regeneration of Diversity

From time to time the biosphere has been beset with mass extinctions of species. The most profound of these crises marked the end of the Ordovician, Permian, Triassic, and Cretaceous periods, but subsidiary crises came during the Late Devonian and at the end of the Eocene epoch.[71] Despite these reductions in diversity, which affected marine communities perhaps more than life on land, the diversity of organisms has increased in a stepwise fashion over the course of the last 600 million years, both locally and globally.[72] This means that extinct species have, on the average, been replaced by newly evolving species. That rediversification can take place, albeit with a substantial time lag, after a crisis as profound as that which occurred at the end of the Cretaceous period, suggests that the biosphere is remarkably resilient, and that the evolutionary potential of species which survived the crisis is great. Much needs to be learned about the nature of surviving species and about the rates of recovery following a crisis, but present evidence suggests that diversification takes place from stocks of biologically unspecialized organisms with a "dull" appearance. Eventually, biotically competent species become reestablished alongside their less competent ancestors. It is important to ask whether the species being destroyed by man can be replaced and, if so, how the new species are likely to differ from the old ones.

The first part of this question deals with the conditions for speciation. In the classical view adopted by Mayr and accepted as most important by a majority of biologists, geographical isolation coupled with a distinctive regime of selection promotes speciation.[73] The habitat fragmentation resulting from human activities should favor this kind of speciation. Whether it has done so is unclear. It would be most interesting to monitor the discontinuously distributed remnants of once continuously distributed species for morphological and biochemical divergence. Studies on introduced species show that such divergence can take place within 50 years. Particularly good examples include the face fly in North America,[74] the mynah in New Zealand,[75] rabbits in Australia,[76] the house sparrow in the Americas and New Zealand,[77] and the mosquito fish in Hawaii.[78]

It may seem paradoxical that conditions which are favorable to extinction are also likely to stimulate speciation. What factors determine whether a population continues to decline to extinction or gives rise to a new species which subsequently expands its range? Because more genetic variation is preserved in an expanding population than in a declining one,[79] the successful incorporation of favorable genetic alterations is more likely in an expanding population than in one which is stable or contracting. Population expansion may be impossible in most instances unless the population's habitat is expanding. Enlargement of refuges, which in man's presence may be far less common than before his rise to world dominance, may thus be an important precondition for the divergence of new geographically widespread species. These ideas are speculative and require critical evaluation.

What would be the characteristics of new species? It is likely that most natural communities, especially those on land, will be confined to relatively small refuges in man's presence. These refuges are in many ways like islands in that they support fewer species than would habitats of similar size that are surrounded by additional appropriate habitat. Moreover, large predators and potentially dominant competitors capable of exercising intense selection on many species are underrepresented on islands and in small refuges. Darwin had already noted that species which are best adapted to the "struggle for life" evolve in areas where the number of locally co-occurring and interacting species is high, and that low diversity is typically associated with a modest development of biological defense.[80] Species likely to evolve on land today thus may be less biologically competent than were their ancestors. Because so many medically interesting chemicals function as defenses against pests and competitors in plants, the species evolving in man's presence may be less rich in these chemicals than were their predecessors. If the preservation and the natural generation of biotically competent species is a goal, then it is of the utmost importance to conserve large tracts of habitat that support a highly diverse biota.

Exploitation of predaceous fishes and invertebrates by man is probably also changing, and perhaps relaxing, selection from biological sources in many unexploited marine species. It is imperative that this hypothesis receive attention from biologists. If exploitation does relax selection, the preservation of traits which evolved under previously more intense selection would be possible only by setting aside substantial refuges in which exploitation by humans is forbidden. Reef fishing, for example, is now so widespread that few tropical locations remain in which the complement of reef fishes approaches the condition prevailing before the arrival of man.

There is another side to this coin. Before man introduced species to oceanic islands, insular biotas consisted of species whose traits were appropriate for existence in an environment of relaxed biological selection, but which would be inappropriate in most continental selective regimes. Accordingly, organisms evolved interesting and potentially important peculiarities which could not evolve on oceanic islands today because introduced species have rendered the selective regime on islands more like that of continents. Examples of these peculiarities include a novel hinge in the jaws of bolyerine boid snakes of the Mascarine Islands,[81] flightlessness in many birds and insects, and tree-like form in normally herbaceous plants of the family Compositae.[82]

What, then, will life with man in the future be like? If the data and speculation in this essay are correct, man may be homogenizing the selectional environment of organisms and inhibiting the evolution of novelty and of biotic competence, even if speciation can still take place. The likelihood of speciation will be small if, as seems likely, refuges from human exploitation are kept small. Economic and population pressures will tend to reduce rather than to increase the size of refuges, and will thus prevent relict populations from expanding and revitalizing. Small refuges are precarious places for the long-term conservation of species because locally bad weather or outbreaks of disease may bring on the global extinction of a species rather than just a local reversible population decline. As long as man does not destroy the world in war, there will be life on earth for the foreseeable future, but it may well lack the diversity, beauty, and biotic competence which it had when man first appeared on the planet.

NOTES

1. Lee Van Valen, "A New Evolutionary Law," *Evolutionary Biology* 1 (1973): 1-18; Geerat J. Vermeij, *Biogeography and Adaptation: Patterns of Marine Life* (Cambridge, Mass.: Harvard University Press, 1978).

2. J. R. Karr, "Avian Extinction on Barro Colorado Island, Panama: A Reassessment," *American Naturalist* 119 (1982) : 220-39.

3. John Terborgh and B. Winter, "Some Causes of Extinction," in *Conservation Biology: An Evolutionary-Ecological Perspective*, edited by M. E. Soule and B. A. Wilcox (Sunderland, Mass.: Sinauer Associates, 1980), pp. 119-33.

4. Paul F. Basch, "Radulae of North American Ancylid Snails. II. Subfamily Neoplanorbinae," *The Nautilus* 75 (1962): 145-49; David H. Stansbery, "Eastern Freshwater Mollusks (I). The Mississippi and St. Lawrence River Systems," *Malacologia* 10 (1970): 9-22.

5. Stansbery, "Eastern Freshwater Mollusks."

6. Geerat J. Vermeij and Alan P. Covich, "Coevolution of Freshwater Gastropods and Their Predators," *American Naturalist* 112 (1978): 833-43.

7. A. Dafni and M. Agami, "Extinct Plants of Israel," *Biological Conservation* 10 (1976): 43-52.

8. G. K. Pregill and S. L. Olson, "Zoogeography of West Indian Vertebrates in Relation to Pleistocene Climatic Cycles," *Annual Review of Ecology and Systematics* 12 (1981): 75-98.

9. S. David Webb, "Extinction-Origination Equilibria in Late Cenozoic Land Mammals of North America," *Evolution* 23 (1969): 668-702.

10. Blake W. Blackwelder, "Late Cenozoic Stages and Molluscan Zones of the

U.S. Middle Atlantic Coastal Plain," *Paleontological Society Memoir 12, Journal of Paleontology* 55 (supplement to no. 5) (1981): 1-34; Steven M. Stanley and Lyle D. Campbell, "Neogene Mass Extinction of Western Atlantic Molluscs," *Nature* 293 (1981): 457-59; Edward J. Petuch, "Geographical Heterochrony: Contemporaneous Coexistence of Neogene and Recent Molluscan Faunas in the Americas," *Palaeogeography, Palaeoclimatology, Palaeoecology* 37 (1982): 277-312.

11. C. O. van Regteren Altena, A. Bloklander, and L. Pouderoyen, "De Fossiele Schelpen van de Nederlandse Stranden en Zeegaten," *Basteria* 18 (1954): 54-64; *Basteria* 19 (1955): 27-34; *Basteria* 20 (1956): 81-90; *Basteria* 21 (1957): 67-73; *Basteria* 25 (1961): 3-6; *Basteria* 28 (1964): 1-9. Geerat J. Vermeij, "Unsuccessful Predation and Evolution," *American Naturalist* 120 (1982): 701-20.

12. L. G. Marshall, "The Great American Interchange: An Invasion Induced Crisis for South American Mammals," in *Biotic Crises in Ecological and Evolutionary Time*, edited by M. H. Nitecki (New York: Academic Press, 1981), pp. 133-229.

13. W. B. King (compiler), "Endangered Birds of the World," in *ICBP Bird Red Data Book* (Washington, D.C.: Smithsonian Institution Press, 1981).

14. Ibid.

15. R. E. Honegger, "List of Amphibians and Reptiles Either Known or Thought to Have Become Extinct Since 1600," *Biological Conservation* 19 (1981): 141-58.

16. Paul R. Ehrlich and Anne Ehrlich, *Extinction: The Causes and Consequences of the Disappearance of Species* (New York: Random House, 1981).

17. Geerat J.Vermeij, David E. Schindel, and Edith Zipser, "Predation Through Geological Time: Evidence From Gastropod Shell Repair," *Science* 214 (1981): 1024-26; Geerat J. Vermeij, "Unsuccesful Predation."

18. R. E. Warner, "The Role of Introduced Diseases in the Extinction of the Endemic Hawaiian Avifauna," *Condor* 70 (1968): 101-20.

19. I.A.E. Atkinson, "A Reassessment of Factors, Particularly *Rattus rattus* L., That Influence the Decline of Endemic Forest Birds in the Hawaiian Islands," *Pacific Science* 31 (1977): 109-33.

20. Ibid.

21. B. E. Coblentz, "The Effects of Feral Goats (*Capra hircus*) on Island Ecosystems," *Biological Conservation* 13 (1978): 279-86.

22. Q.C.B. Cronk, "Extinction and Survival in the Endemic Vascular Flora of Ascension Island," *Biological Conservation* 13 (1980): 207-19.

23. W. J. Christie, "Changes in the Fish Species Composition of the Great Lakes," *Journal of the Fisheries Research Board of Canada* 31 (1974): 827-54.

24. J. M. Diamond and C. R. Veitsch, "Extinctions and Introductions in the New Zealand Avifauna: Cause and Effect?" *Science* 211 (1981): 499-501.

25. Ibid.

26. Daniel S. Simberloff, "Community Effects of Introduced Species," in Nitecki, *Biotic Crises*, pp. 53-81.

45

27. Charles S. Elton, *The Ecology of Invasions by Animals and Plants* (London: Methuen, 1958).
28. Vermeij, *Biogeography*.
29. Margaret S. Race, "Field Ecology and Natural History of *Cerithidea Californica*," *Veliger* 24 (1981): 18-27.
30. Vermeij, *Biogeography*; J. W. Ropes, "The Feeding Habits of the Green Crab, *Carcinus maenas* (L.)," *Fisheries Bulletin* 67 (1968): 187-203; M. D. Bertness, "Habitat and Community Modification by an Introduced Herbivorous Snail," *Ecology* 65 (1984): 370-81; Geerat J. Vermeij, "Environmental Change and the Evolutionary History of the Periwinkle *Littorina littorea* in North America," *Evolution* 36 (1982): 561-80.
31. Vermeij, *Biogeography*.
32. Peter B. Moyle, "Fish Introductions in California: History and Impact on Native Fishes," *Biological Conservation* 9 (1976): 101-18.
33. Ibid.
34. Thomas M. Zaret and Robert T. Paine, "Species Introduction in a Tropical Lake," *Science* 182 (1973): 449-55; Karr, "Avian Extinction."
35. A. L. Barash and Z. Danin, "Additions to the Knowledge of Indo-Pacific Mollusca in the Mediterranean," *Conchiglie* 13 (1977): 85-116; A. Ben-Tuvia, "Immigration of Fishes Through the Suez Canal," *Fisheries Bulletin* 76 (1978): 249-55; D. F. Por, "One Hundred Years of Suez Canal—A Century of Lessepsian Migration: Restrospect and Viewpoints," *Systematic Zoology* 20 (1971): 138-59; Vermeij, *Biogeography*.
36. M. Ben-Yami and T. Glaser, "The Invasion of *Saurida undosquamis* (Richardson) into the Levant Basin—An Example of Biologic Effects of Interoceanic Canals," *Fisheries Bulletin* 72 (1974): 359-73.
37. Y. Achituv, "On the Distribution and Variability of the Indo-Pacific Sea Star *Asterina wega* (Echinodermata: Asteroidea) in the Mediterranean Sea," *Marine Biology* 18 (1973): 333-36.
38. Vermeij, *Biogeography*; Geerat J. Vermeij and John D. Currey, "Geographical Variation in the Strength of Thaisid Snail Shells," *Biological Bulletin* 158 (1980): 383-89; Mark D. Bertness, "Shell Utilization, Predation Pressure, and Thermal Stress in Panamanian Hermit Crabs: An Interoceanic Comparison," *Journal of Experimental Marine Biology and Ecology* 64 (1982): 159-87.
39. Peter W. Glynn, "Some Physical and Biological Determinants of Coral Community Structure in the Eastern Pacific," *Ecological Monographs* 46 (1976): 431-56; James W. Porter, "Community Structure of Coral Reefs on Opposite Sides of the Isthmus of Panama," *Science* 186 (1974): 543-45.
40. Marshall, "Great American Interchange."
41. Friedrich Strauch, "Die Thule-Lanbrucke als Wanderweg und Faunecheide Zwischen Atlantik und Skandik im Tertiar," *Geologische Rundschau* 60 (1970): 381-417.
42. Vermeij, *Biogeography*.

43. John C. Briggs, "Operation of Zoogeographic Barriers," *Systematic Zoology* 23 (1974): 248-56.
44. Terborgh and Winter, "Some Causes of Extinction."
45. Ibid.
46. Karr, "Avian Extinction."
47. John Faaborg, "Metabolic Rates, Resources and the Occurrence of Non-passerines in Terrestrial Avian Communities," *American Naturalist* 111 (1977): 903-16; F. Harvey Pough, "The Advantages of Ectothermy for Tetrapods," *American Naturalist* 115 (1980): 92-112.
48. S. J. Wright, "Competition between Insectivorous Lizards and Birds in Central Panama," *American Zoologist* 19 (1979): 1145-56.
49. Peter Raven, "A Suggestion Concerning the Cretaceous Rise to Dominance of the Angiosperms," *Evolution* 31 (1977): 451-52; D. M. Mulcahy, "The Rise of the Angiosperms: A Genecological Factor," *Science* 206 (1979): 20-23.
50. Vermeij, *Biogeography*; Mulcahy, "Rise of Angiosperms"; J.R.G. Turner, "Adaptation and Evolution in Heliconius: A Defense of Neo-Darwinism," *Annual Review of Ecology and Systematics* 12 (1981): 99-121.
51. R. Tucker Abbott, "Eastern Marine Mollusks," *Malacologia* 10 (1970): 47-49; A. Myra Keen, "Western Marine Mollusks," *Malacologia* 10 (1970): 51-53.
52. Erik Rasmussen, "Systematics and Ecology of the Isefjord Marine Fauna (Denmark)," *Ophelia* 11 (1973): 1-495.
53. Ibid.
54. Vermeij, *Biogeography*.
55. Ibid.
56. Karr, "Avian Extinction."
57. Vermeij, *Biogeography*.
58. Harry S. Ladd, "Cenozoic Fossil Mollusks From Western Pacific Islands: Chitons and Gastropods (Haliotidae through Adeorbidae)," *U.S. Geological Survey Professional Paper 531* (Washington, D.C.: U.S. Geological Survey, 1966), pp. 1-98; "Gastropods (Turritellidae through Strombidae)," *U.S. Geological Survey Professional Paper 532* (Washington, D.C.: U.S. Geological Survey 1972), pp. 1-79; "Gastropods (Eratoidae through Harpidae)," *U.S. Geological Survey Professional Paper 533* (Washington, D.C.: U.S. Geological Survey, 1977), pp. 1-84.
59. Geerat J. Vermeij and E. J. Petuch, "Differential Extinction in Tropical American Molluscs: Endemism, Architecture, and the Panama Land Bridge," *Malacologia* 28 (in press).
60. Vermeij, *Biogeography*.
61. R. T. Balaker, "Tetrapod Mass Extinctions—A Model of the Regulation of Speciation Rates and Immigration by Cycles of Topographic Diversity," in A. Hallam, editor, *Patterns of Evolution, As Illustrated by the Fossil Record* (Amsterdam: Elsevier, 1977), pp. 439-68.
62. L. F. Lowry and J. S. Pearse, "Abalones and Sea Urchins in an Area Inhabited

by Sea Otters," *Marine Biology* 23 (1973): 213-19; J. A. Estes, N. S. Smith, and J. F. Palmisano, "Sea Otter Predation and Community Organization in the Western Aleutian Islands," *Ecology* 59 (1978): 822-33; C. A. Simenstad, J. A. Estes, and K. W. Kenyon, "Aleuts, Sea Otters, and Alternate Stable State Communities," *Science* 200 (1978): 403-11.

63. Paul K. Dayton, "Experimental Studies of Algal Canopy Interactions in a Sea Otter-Dominated Kelp Community at Amchitka Island, Alaska," *Fisheries Bulletin* 73 (1975): 230-37.

64. M. J. Tegner, "Multispecies Considerations of Resource Management in Southern California Kelp Beds," *Canadian Technical Report of Fisheries and Aquatic Sciences* 954 (1980): 125-43; W. K. Wharton and K. H. Mann, "Relationship between Destructive Grazing by the Sea Urchin, *Strongylocentrotus droebachiensis*, and the Abundance of American Lobster, *Homarus americanus*, on the Atlantic Coast of Nova Scotia," *Canadian Journal of Fisheries and Aquatic Sciences* 38 (1981): 1339-49.

65. Stanley A. Temple, "Plant-Animal Mutualism: Coevolution with Dodo Leads to Near Extinction of Plant," *Science* 197 (1977): 885-86.

66. Daniel H. Janzen and Paul S. Martin, "Neotropical Anachronisms: The Fruits the Gomphotheres Ate," *Science* 215 (1982): 19-27.

67. Ibid.

68. Ehrlich and Ehrlich, *Extinction*.

69. K. J. Hsu, Q. He, J. A. McKenzie, H. Weissert, K. Perch-Nielsen, H. Oberhansli, K. Kelts, J. LaBrecque, L. Tauxe, U. Krahenbuhl, S. F. Percival, Jr., R. Wright, A. M. Karpoff, N. Petersen, P. Tucker, R. Z. Poore, A. M. Gombos, K. Pisciotto, M. F. Carmann, Jr., and E. Schreiber, "Mass Mortality and Its Environmental and Evolutionary Consequences," *Science* 216 (1982): 249-56.

70. T.J.M. Schopf, *Paleoceanography* (Cambridge, Mass.: Harvard University Press, 1980).

71. David M. Raup and John J. Sepkoski, Jr., "Mass Extinction in the Marine Fossil Record," *Science* 215 (1982): 1501-1503.

72. Richard K. Bambach, "Species Richness in Marine Benthic Habitats Through the Phanerozoic," *Paleobiology* 3 (1977): 152-67; John J. Sepkoski, Jr., Richard Bambach, David M. Raup, and James W. Valentine, "Phanerozoic Marine Diversity and the Fossil Record," *Nature* 291 (1981): 435-37.

73. Ernst Mayr, *Animal Species and Evolution* (Cambridge, Mass.: Harvard University Press, 1963).

74. E. H. Bryant, H. van Dijk, and W. van Delden, "Genetic Variability of the Face Fly, *Musca autumnalis* de Geer in Relation to a Population Bottleneck," *Evolution* 35 (1981): 872-81.

75. A. J. Baker and A. Moeed, "Evolution in the Introduced New Zealand Populations of the Common Myna, *Acridotheres tristic* (Aves: Sturnidae)," *Canadian Journal of Zoology* 57 (1979): 570-84.

76. B. J. Richardson, "Ecological Genetics of the Wild Rabbit in Australia, III. Comparison of the Microgeographical Distribution of Alleles in Two Dif-

ferent Environments," *Australian Journal of Biological Sciences* 33 (1980): 385-91; B. J. Richardson, P. M. Rogers and G. M. Hewitt, "Ecological Variation in British, French and Australian Rabbits and the Geographical Distribution of the Variation in Australia," *Australian Journal of Biological Sciences* 33 (1980): 371-83.

77. A. J. Baker, "Morphometric Differentiation in New Zealand Populations of the House Sparrow (*Passer domesticus*)," *Evolution* 34 (1980): 638-53; Richard F. Johnston and Robert K. Selander, "House Sparrows: Rapid Evolution of Races in North America," *Science* 144 (1964): 548-50; Richard F. Johnston and Robert K. Selander, "Evolution in the House Sparrow, II. Adaptive Differentiation in North American Populations," *Evolution* 25 (1971): 1-28; Richard F. Johnston and Robert K. Selander, "Evolution in the House Sparrow, III. Variation in Size and Sexual Dimorphism in Europe and North and South America," *American Naturalist* 107 (1973): 373-90.

78. Steven C. Stearns and Richard D. Sage, "Maladaptation in a Marginal Population of the Mosquito Fish, *Gambusia affinis*," *Evolution* 34 (1980): 65-75.

79. E. B. Ford, *Ecological Genetics*, second ed. (London: Methuen, 1965); C. J. Krebs, M. S. Gaines, B. L. Keller, J. H. Myers, and R. H. Tamarin, "Population Cycles in Small Rodents," *Science* 179 (1973): 35-41.

80. Charles Darwin, *The Origin of Species by Natural Selection or the Preservation of Favored Races in the Struggle for Life*, sixth ed. (New York: Colliers, 1872).

81. T. H. Frazzetta, "From Hopeful Monsters to Bolyerine Snakes?" *American Naturalist* 104 (1970): 55-71.

82. S. Carlquist, "The Biota of Long-Distance Dispersal II. Loss of Dispersity in Pacific Compositae," *Evolution* 20 (1966): 30-48.

3

Social and Perceptual Factors
in the Preservation of
Animal Species

STEPHEN R. KELLERT

I. Introduction

The development of a compelling rationale and an effective strategy for protecting endangered species will require an increasing recognition that most contemporary extinction problems are largely the result of socio-economic and political forces. Awareness of the socioeconomic basis of the problem should be a strong factor governing recovery and preservation efforts.

The historical experience of Hawaii's extraordinary avifauna is a classic example of the importance of these social forces.[1] Some researchers have suggested that if Darwin had traveled to the more geographically isolated Hawaiian, rather than Galapagos, Islands, he would have encountered in the Hawaiian honeycreepers a family of birds even better suited than the Galapagos finches to illustrate his evolutionary theories. The Hawaiian Islands, however, never achieved the same biological visibility, and, in the years since the islands' discovery, most of their endemic birdlife has become extinct or endangered. In the great majority of cases, human-related factors were the primary causes for the decline of these species. Critical factors included overhunting for meat and feathers, agriculture-related and grazing-related habitat destruction, the impacts of feral herbivores and carnivores, predation by domestic pets, competition for food with introduced birds and insects, overambitious animal damage control, egg-eating by exotic mammals and non-native birds, extensive forest clearing, excessive tourist development, introduction of avian malaria, overcollecting by private hobbyists, and even the creation of game habitat by state wildlife agencies and the introduction of non-native birds by frustrated birdwatchers. While Hawaii's endemic birdlife was biologically vulnerable to these impacts, sociopolitical and economic forces created the contextual basis for this decline, and only the mitigation of these social factors can counter the continuing drift toward extinction of the remaining endangered species.

Although this assertion may seem obvious, most endangered species preservation efforts pay scant attention to human social factors and perceptions. The typical endangered species program tends to be preoccupied with biological assessments and biological solutions. Perhaps this bias reflects the training of most wildlife professionals in the ecological rather than social sciences and the hope for a technological "quick-fix" solution to the problem. It may also reflect the enormous complexity of the issue when viewed socioeconomically and the political risks associated with any perspective that suggests altering societal institutions and perceptions as an appropriate remedial response.

To suggest that the causes of a problem are inextricably woven into the fabric of human society implicitly assumes the need to assign fundamental social and perceptual forces a central role in devising solutions. A broad social perspective is further indicated by the prospect, as some have suggested, of hundreds of thousands of species extinctions during the next quarter to half century if contemporary rates of habitat destruction continue, particularly in the tropical moist forests.[2] The possibility of this species loss emphasizes the problem not as a case of one or two animals whose rarity engages our altruistic capacities, nor as the luxurious indulgence of an affluent society, nor as the esoteric preoccupation of academic scientists. Instead, it suggests that we need to engage in a fundamental reassessment of the relationship between human society and the natural world. This viewpoint implies, in other words, that the issue is our ability to manage the planet's biological resources not in the service of compassion, kindness, and altruism, but in the interest of human economic and social well-being.

The objective of this chapter is to review various social and perceptual factors contributing to the contemporary species extinction and endangerment problem. If my central thesis, that the endangered species problem results mainly from socioeconomic and political forces, is correct, then a consideration of social and perceptual factors is essential to an understanding of the problem. To the extent that attitudinal, social, and economic forces characteristic of contemporary life endanger species, an understanding of these forces is a necessary prerequisite to proposing strategies for mitigating their impact.

Without attempting to be exhaustive, I will survey a number of especially salient aspects of important socioeconomic issues relating to the endangered species problem and, in particular, stress the critical significance of human perceptions and attitudes as a force that must be reckoned with in any potentially effective remedial program. Empirical data concerning American attitudes toward and concern for protecting rare and imperiled species will be presented. The primary intent of this chapter, then, is not to advocate solutions, but to delineate the major contours of the socioeconomic character of the contemporary extinction crisis. This effort should provide a partial basis for defining a social research and policy agenda facilitating a more meaningful search for solutions.

I begin by arguing that the values and benefits people associate with endangered species have been insufficiently identified and empirically assessed in situations involving conflicts between various human activities and species preservation. It will be noted, however, that these social valuations vary considerably when comparing invertebrate with vertebrate species. The discussion will first focus on the attitudes of people in industrially advanced, primarily Western societies. Since the projected large-scale species loss during the next quarter-century will occur primarily among invertebrate species in the tropical, less developed countries, I will then turn to a review of social factors in industrially undeveloped countries. Finally, I will conclude by listing some of the strategies for preserving species implied by the previous discussion.

II. Attitudes toward Endangered Species in Industrially Developed Countries

One of the most fundamental deficiencies in nearly all assessments of the need for expending scarce resources on endangered species is an insufficient appreciation of people's perceptions of wildlife, as well as an undervaluation of the benefits derived from creatures in more or less natural surroundings. This problem usually manifests itself in two ways: an incomplete specification of the full range of values associated with biological diversity, and a lack of empirical, numerical, and commensurable assessments of these values. These inadequacies usually result in decisions which, by default, place greater emphasis on monetary parameters, typically serving the interests of economic development and resource exploitation.

The first task, then, is to identify those values and benefits people associate with and derive from the existence of wildlife. For developed countries at least seven discrete values can be listed and, to a limited extent, quantitatively assessed.[3] These include:

(1) *Naturalistic/outdoor recreational value*: There is enjoyment to be

gained from direct contact with wildlife while taking part in some outdoor activity such as camping, backpacking, canoeing, etc. The opportunity to observe rare species (e.g., bighorn sheep, grizzly bear, bald eagle) has often been cited as a major part of a satisfactory backcountry experience.[4]

(2) *Ecological value*: Particular species and associated habitats are important to the well-being and continuity of interrelated flora and fauna, as well as to the maintenance of basic hydrological, soil, and other biogeochemical processes.

(3) *Moral or existence value*: It is possible to believe that all species have inherent rights and spiritual importance. Some writers have emphasized the right to survival possessed by any species, and the duties consequently imposed on humans to protect and preserve all life forms.[5] Others have noted the value of knowing that a rare species such as the blue whale exists, even though most people will never see one.

(4) *Scientific value*: All species have actual or potential value for advancing human knowledge and understanding of the natural world.

(5) *Aesthetic value*: Many plants and animals are physically attractive. Some rare animals in particular (such as the great white heron, mountain goat, or giant panda) have been recognized as possessing great beauty and other aesthetic qualities.

(6) *Utilitarian value*: Natural populations of organisms are potential sources of material benefit to human society, especially in the realms of medicine, agriculture, and industry.[6]

(7) *Cultural, symbolic, and historic value*: Both animals and plants may function as expressions of group identity or social experiences and be the objects of specialized attachments. Related to this is what might be called *humanistic value*, i.e., strong affection for individual animals. Certain rare species (e.g., elephant, tiger), because of their anthropomorphic and historic significance, have been the recipients of strong personal and symbolic meanings.

As previously suggested, the relevance and range of these values have rarely been identified or adequately assessed in most conflict situations involving choices between species preservation and diverse human activities that might imperil these species. An example might illustrate this point—the now infamous Tellico Dam/snail darter controversy.

The snail darter is 2-3 inches long, a member of the perch family. Of approximately 130 darter species, 85-90 occur in the state of Tennessee, with some 45 species found in the Tennessee River area. The snail darter was thought to be endangered by the construction of the Tellico Dam, because the dam's impact on the river current and stream bottom were likely to result in a significant decrease in the darter's food supply and spawning areas.

A review of two documents—the Tennessee Valley Authority's cost-benefit analysis and the General Accounting Office's report to Congress—reveals the difficulties involved in assessing the value of an endangered species and its habitat.[7] The GAO report, for example, ignored the moral, ecological, aesthetic, scientific, and other noncommodity and amenity values associated with the area or the snail darter. Instead, this "cost-benefit" assessment focused almost entirely on dollar values associated with land and construction costs, bridges, dams, reservoirs, spillways, recreational expenditures, flood control benefits, navigational costs, power benefits, hunting and fishing expenditures, and water supply opportunities. Cultural and historical values were discussed to a very limited extent. The report was clearly biased toward commodity values and, without question, monetary measures did not merely dominate but overwhelmed the assessment.

The Endangered Species Committee recommended the dam not be completed, primarily on the basis of its quantitative assessment of construction costs, dam benefits, and flooding impacts. Nevertheless, the dam did receive budgetary approval by Congress, which concluded that the known benefits clearly outweighed vague values associated with an obscure and rare fish species, local agricultural interests, and native American burial grounds. The final Congressional debate stressed the power-related benefits of the project which, given the energy-crisis mentality of the time, were more persuasive than the arguments emphasizing recreational and industrial potential originally put forward by TVA. This difference in public susceptibility to varying rationalizations was additionally reflected in the results of a national study of over three thousand randomly selected citizens.[8] In answering a question simulating the Tellico Dam–snail darter case, respondents expressed a willingness to forgo creating a lake for recreational purposes or diverting water for industrial development if these projects might endanger an obscure fish species (Table 1). On the other hand, if the project were going to produce hydroelectric, agricultural, or drinking water improvements, the public overwhelmingly supported the activity regardless of the impact on the rare fish species.

The snail darter controversy suggests that, first, a bias often exists in the minds of most analysts, the general public, and legislative decision-makers toward quantifiable material benefits, especially if measurable in dollar terms and related to critical human needs (e.g., food, energy, jobs). Second, and perhaps more important, since species-related values are often incompletely specified and hard to measure, the tendency is grossly to understate the risks involved when living resources are impaired or destroyed.

Table I

Public Approval and Disapproval of Diverse Water Uses That Endanger Fish Species

Various kinds of fish have been threatened with extinction because of dams, canals, and other water projects. Please indicate if you would approve of the following water uses if they were to endanger a species of fish.

	Strongly Approve	Approve	Slightly Approve	Slightly Dis- approve	Disap- prove	Strongly Dis- approve
A. Water diverted to cool *industrial plant* machinery*	(75) 3.1	(532) 21.7	(569) 23.2	(422) 17.2	(543) 22.1	(175) 7.1
Total			48			46
B. Water dammed to provide *hydro- electric energy power*	(190) 7.7	(937) 38.2	(631) 25.7	(250) 10.2	(248) 10.1	(80) 3.3
Total			72			24
C. Water diverted to increase *human drinking* supplies	(456) 18.6	(1267) 51.6	(405) 16.5	(121) 4.9	(103) 4.2	(39) 1.6
Total			87			11
D. Water dammed to make a lake for *recreational use*	(88) 3.6	(432) 17.6	(444) 18.1	(381) 15.5	(702) 28.6	(327) 13.3
Total			39			57
E. Water diverted to irrigate *agri- cultural crops*	(311) 12.7	(1157) 47.1	(576) 23.4	(166) 6.8	(127) 5.2	(47) 1.9
Total			83			14

*Note difference between approve and disapprove on this water use is not significant with $Z = .75$, $P = .45$. All other water use differences between approve and disapprove are highly significant, $P = \leq .0001$.

We need, therefore, to be far more precise about the various benefits people derive from biological diversity, as well as the actual values they associate with it. Until this is accomplished, we will lack an adequate basis for making trade-offs between species preservation and socioeconomic development. There is some evidence that the values people attach to endangered species are far greater than we might at first realize. Studies of the American public have revealed that in a variety of socioeconomic areas—including forestry, industry, housing, mineral development, agriculture, and energy development—a significant majority are willing to forgo economic gains that would adversely affect wildlife (for example, Tables 2 and 3).[9]

The challenge is to identify all species values and to measure these values in an empirical and standardized fashion. These numerical standards should be applicable across diverse settings and make it possible to assign relative weights to the many factors in land-use decisions. This empirical emphasis may not adequately express the intangible, spiritual benefits derived from diverse forms of life—what Aldo Leopold once metaphorically described as "the sound of cranes signifying the trumpet in the orchestra of evolution."[10] But in the context of painful trade-offs between species preservation and socioeconomic development, few practical alternatives may exist for incorporating concern for wildlife into prevailing decision-making procedures.

The primary need is to ensure that considerations of species preservation are not perceived and treated apart from fundamental socioeconomic decisions. Indeed, the exclusion of such environmental assessments from most societal evaluations may historically have been the most significant factor in the process leading to the decline and endangerment of many species. To regard any economic system as environmentally separate, independent, and superior is, in other words, to invite species degradation and decline.

Our ability to consider human attitudes and valuations of species is seriously limited, however, by three characteristics of the contemporary extinction crisis. First, as previously suggested, present rates of habitat destruction, particularly in the tropics, imperil an extraordinary number of species, perhaps in the hundreds of thousands. Second, most of these endangered species are invertebrates, mainly insects. Third, this species loss will primarily occur in the less developed countries, characterized by very low per capita incomes, economic dependence on natural resources for hard currency needs, and often serious overpopulation pressures.

The particular problem arising from the projected scale of species loss is that it renders difficult, and perhaps inappropriate, valuations of species

Table 2
**Public Attitudes toward Modifying Energy Projects to Protect
Selected Endangered Species**

A recent law passed to protect endangered species may result in changing some energy development projects at greater cost. As a result, it has been suggested that endangered species protection be limited only to certain animals and plants. Which of the following endangered species would you favor protecting, *even if it resulted in higher costs for an energy development project?*

	Strongly Favor	Favor	Slightly Favor	Slightly Oppose	Oppose	Strongly Oppose	N.O.
A. A *butterfly*, such as the Silverspot Butterfly	(232) 9.5	(717) 29.2	(612) 25.0	(228) 9.3	(371) 15.1	(91) 3.7	(201) 8.2
Total			64			28	
B. The Eastern *Mountain Lion*	(409) 16.7	(880) 35.9	(504) 20.6	(161) 6.6	(292) 11.9	(69) 2.8	(138) 5.6
Total			73			21	
C. A *fish*, such as the Agassiz Trout	(293) 11.9	(901) 36.7	(550) 22.4	(178) 7.2	(263) 10.7	(64) 2.6	(201) 8.2
Total			71			21	
D. A *spider*, such as the Kauai Wolf Spider	(115) 4.7	(292) 11.9	(428) 17.4	(338) 13.8	(710) 28.9	(343) 14.0	(226) 9.2
Total			34			57	
E. The *American Crocodile*	(325) 13.2	(859) 35.0	(526) 21.5	(179) 7.3	(331) 13.5	(104) 4.2	(127) 5.2
Total			70			25	
F. A *plant*, such as the Furbish Lousewort*	(140) 5.7	(510) 20.8	(519) 21.1	(266) 10.8	(438) 17.8	(127) 5.2	(451) 18.4
Total			48			34	
G. A *snake*, such as the Eastern Indigo Snake**	(147) 6.0	(471) 19.2	(447) 18.2	(264) 10.7	(622) 25.3	(304) 12.4	(197) 8.0
Total			43			49	
H. A *bird*, such as the Bald Eagle	(1078) 43.9	(907) 37.0	(199) 8.1	(63) 2.6	(106) 4.3	(29) 1.2	(70) 2.8
Total			89			8	

* Favor versus oppose differences were significant: $Z = 7.56$, $P = \leq .0001$
** Favor versus oppose differences were significant: $Z = 2.63$, $P = < .0009$

Table 3
Public Attitudes toward Habitat Preservation Questions

Natural resources must be developed even if the loss of wilderness results in much smaller wildlife populations.

Strongly Agree	Agree	Slightly Agree	Slightly Disagree	Disagree	Strongly Disagree
(69)	(498)	(515)	(431)	(611)	(198)
2.8	20.3	21.0	17.5	24.9	8.1
Total		44			51

Z = 3.28, P = .001

I approve of building on marshes that ducks and other nonendangered wildlife use if the marshes are needed for housing development.

Strongly Agree	Agree	Slightly Agree	Slightly Disagree	Disagree	Strongly Disagree
(65)	(486)	(405)	(309)	(715)	(365)
2.6	19.8	16.5	12.6	29.1	14.9
Total		39			57

Z = 8.94, P = ≤ .0001

Cutting trees for lumber and paper should be done in ways that help wildlife even if this results in higher lumber prices.

Strongly Agree	Agree	Slightly Agree	Slightly Disagree	Disagree	Strongly Disagree
(343)	(1031)	(496)	(216)	(230)	(51)
14.0	42.0	20.2	8.8	9.4	2.1
Total		76			20

Z = 28.22, P = ≤ .0001

on a single-case basis.[11] What could be a reasonable evaluation of an individual endangered animal might be impossible to conduct when dealing with hundreds, even thousands, of rare and threatened species. Some have suggested, instead, that the only meaningful response to a challenge of this size is an "ecosystem" approach focusing attention on preserving large areas of habitat essential to the survival of many imperiled species.

An ecosystem approach may seem reasonable, but it will be necessary to convince the public and its political decision-makers that the problem warrants this degree of land protection. Species extinctions on a large scale, primarily affecting invertebrates—many unknown and found largely in the tropics—constitute a nearly "invisible" class of environmental catastrophe. If people are to be persuaded of the necessity of saving species, they must first "see" and relate to the problem. From the perspective of what is phenomenologically important to the ordinary person, a single-species approach emphasizing biologically familiar animals may be an emotional, perceptual, and even ethical necessity.

Fortunately, most animals are hierarchically organized in food chains, with many of the largest, most affectively and cognitively meaningful animals at the top of these energy pyramids. It has proved possible to emphasize the value of these animals, and in this way gain the necessary public support to establish habitat preserves. In the process, all species involved in the food webs of the target species will be affected; if the particular endangered species designated for protection is judiciously selected, many other forms of life will also be protected. For example, attempts to preserve Madagascar's extraordinary endemic fauna could be greatly enhanced by focusing attention on the island's relatively popular and endangered lemur and loris species.

The second major impediment to assigning values to endangered species is that most of the imperiled species are invertebrates. It becomes necessary, as a consequence, to consider public perceptions of the so-called "lower" life forms, particularly insects. Lovejoy has somewhat facetiously labeled this public bias against invertebrates "vertebrate chauvinism," reflecting the fact that most people tend to view these creatures with a mixture of anxiety and disdain.[12] This antipathy probably has some of its roots in atavistic and cultural norms, which typically associate invertebrates, especially insects, with irrational impulses, an inability to experience pain, and the absence of intelligence.

Only a small minority of people possess much concern or empathy for the plight of endangered invertebrates. Unfortunately, the fate of many of these creatures will, nonetheless, depend on the peculiar collection of sentiments, attitudes, and beliefs manifested toward the spineless kingdom. Thomas Allen reflected this view in noting:

> The study of vanishing wildlife is necessarily the study of . . . [man's perceptions of animals]. What we fear, what we hope, and what we admire in animals will inevitably determine their fate. . . . We have long portrayed ourselves as members of a species that reigns from the summit of life. Around us are our fellow mammals, great and small. The birds soar and flutter for our delight. Fishes, especially

those that rise to our hooks, are considered acceptable. Snakes, frogs, alligators? Yes, reptiles and amphibians are there, but most of them figure as villains in our myths. Far below, disappearing from sight are the invertebrates: insects and worms, sponges and jellyfish, snails and squid. We may enjoy eating a lobster or seeing a butterfly, but from our lofty view we give little heed to the creatures we call lower life forms.[13]

As Allen intimates, invertebrates are viewed somewhat more positively when they possess some aesthetic or utilitarian value. It is much easier to enlist support for the protection of an endangered butterfly or bee species than the preservation of a rare spider or worm. Using the typology of wildlife-related values previously described, few invertebrates appear to possess much recreational, naturalistic, moral, or humanistic value for the ordinary person.

For example, invertebrates are rarely given moral consideration. For most people, moral worth is based on presumptions regarding the animal's capacity for experiencing pain or thought—i.e., as an individual capable of eliciting empathy. A moral perspective on animals is related to a concern for the creature's presumed ability to suffer, an attribute rarely associated with invertebrates. Because invertebrates are not perceived as experiencing beings, most people feel no particular obligation to safeguard the welfare of these animals. This perspective emphasizes the dominant presumption that insects are devoid of a mental life.

Although invertebrates are regarded as of little moral or humanistic worth, most biologists see them as possessing considerable ecological and scientific value, and emphasize the importance of their preservation on these grounds. From an ecological perspective, the contributions of many invertebrates to nutrient cycling, pollination, pest control, and seed dispersal have been cited as indicative of their worth in helping to maintain various food and energy chains. If we look at a living community as a whole, the plight of the individual animal is typically secondary, subordinated to a more population-level focus on species in relation to one another and the natural environment. Unfortunately, the average person only dimly recognizes these ecological and scientific benefits.

Prevailing perceptions of invertebrates emphasize the importance of utilitarian and aesthetic rationalizations in promoting the need for their protection. This pragmatic emphasis is referred to by Norton as "anthropocentric utilitarianism,"[14] which Martin describes in this fashion:

A reason frequently given for preserving [invertebrate] species and ecosystems is their unique . . . properties. Sometimes these items are held to be valuable because their properties are of scientific interest, but many times their importance lies in the beneficial effect the object has on the human environment. In either case, objects with these properties are valuable because of their long-term effect on social utility. Such arguments are properly seen as being consistent with economic utilitarianism.[15]

Because of the importance of utilitarian arguments in invertebrate species preservation, the recent books by Myers and the Erhlichs, and the 1981 Congressional testimony of Raven and Wilson, are particularly significant for documenting the practical benefits of invertebrates for advances in medicine, agriculture, and industry.[16] Peter Raven, for example, noted in his testimony on the utilitarian value of a particular plant group:

> The Antioch Dunes Evening Primrose was federally listed [as an endangered species] because it occurred . . . where there were two species of endangered butterflies. . . . If the butterflies were not there, development of the dunes . . . might have continued. . . . The reason these neglected plants proved of interest . . . was the discovery that the oil in their seeds is one of the only two known rich natural sources of a nutrient called gamma-linolenic acid. . . . Oil derived from the seed . . . may prove to play an essential role in helping us to avoid coronary heart disease and to cure such diseases as eczema, and arthritis. . . . Would the loss of the species have been progress and, if so, for whom would it have been progress? It surely brings to mind a memorable sentence. . . . "We have not inherited the Earth from our parents, we have borrowed it from our children."[17]

Edward Wilson also noted at the same Congressional hearing:

> In reflecting on the preservation of species and genetic diversity, it is worth remembering that . . . scientists have documented the vast opportunities offered by species variation for the development of new crops, drugs, and renewable energy sources. . . . The great German zoologist Karl von Frisch once said that the honeybee is like a magic well; the more you draw from it, the more there is to draw. And so it is with any species, which is a unique configuration of genes assembled over thousands of millions of years, possessing its own biology, mysteries, and still untested uses for mankind.[18]

Since perceptions of the value of nonhuman life forms vary so much, we are led to the extremely difficult issue of the need for a "perceptual calculus" in assigning species preservation priorities. This possibility is anathema to most scientists, although the public apparently has little trouble in making such judgments. An examination of the public's willingness to protect various endangered species has suggested that greatest concern is typically expressed for creatures that are large, aesthetically attractive, phylogenetically similar to human beings, and regarded as possessing capacities for feeling, thought, and pain. Shepard refers to these species as the "phenomenologically" significant animals.[19]

A further assessment, by Burghardt and Herzog, suggests a number of other underlying reasons for these species preferences.[20] They identify four major types: factors associated with human benefits, anthropomorphic factors, ecological factors, and psychological factors. Human benefit factors include an animal's capacity to provide food, clothing, recreation, and companionship, and also its potential for adversely affecting crops, property, and health. Important anthropomorphic factors include presumptions about the animal's capacity for experiencing pain, its cuteness, size, phylogenetic similarity to people, humanoid appearance, mental similarity, and gory or disgusting habits. Ecological factors, although not typically important determinants of public perception, include an animal's rarity and its contribution to diversity and ecological balance. Important psychological factors include the animal's aesthetic characteristics, spiritual and religious associations, symbolic relationship with the wild, and habituating capacity and behavioral plasticity.

These factors, as well as the wildlife-related values already described, outline the perceptual categories people typically employ in deciding which species are worthy of preservation. This type of calculation probably results in a far greater emphasis on the so-called higher than lower life forms. This likelihood usually produces considerable discomfort and is opposed by most biologists as it represents a somewhat antiscientific decision-making process. We need to recognize, however, that most endangered species are at risk not because of their biological inadequacies, but because of a variety of human social, psychological, and cultural factors.

III. SOCIAL FACTORS IN DEVELOPING COUNTRIES

The third impediment to the development of an effective approach to the current species preservation crisis is the fact that a great many of the species with which we are concerned live in the developing world, particularly the countries of the moist tropical forest regions. Among the many difficulties confronting these countries, which have both contributed to and prevented a meaningful solution of the problem of extinction, are very low per capita incomes, overwhelming needs for hard currency, a dependence on raw materials for trade, overpopulation, and lack of an organized and effective wildlife management infrastructure.

It has been noted that many of the less developed countries lack a "conservation ethic." The comparative absence of an animal protectionist philosophy should not suggest, however, indifference to or disrespect for wildlife. It reflects more the absence of relatively affluent sporting and animal welfare groups which, in the Western nations, have historically evolved into constituencies especially concerned with the protection of birds, fur-bearing animals, and game fish.

Generally we find that people in the less developed countries view animals in two ways. The first has its roots in subsistence economies which tend to emphasize the practical and material values of animals. The other perspective often entails a sense of awe and respect for animals but, typically, in a very abstract, idealized, mystical, and sacral context. This latter perspective is often reflected in religious, mythic, and philosophical thought, but usually exerts only minor influence on personal treatment or public management of wild animals or natural habitats.

A great deal of research is needed, as a consequence, on attitudes toward and uses of wildlife in developing countries in the areas of folklore, religion, animal domestication, food taboos, decorative arts, and related aspects of sacral and symbolic relationships to animals. The utilitarian and mystical perspectives of animals should thus receive special consideration in any species preservation effort promulgated in the non-Western, underdeveloped countries.

The most serious obstacle to the evolution of an effective endangered species strategy for the less developed nations is their depressed socioeconomic condition. Widespread poverty and the need to exploit natural resources as a major source of national income appear, at least superficially, to render efforts to preserve species an impractical luxury. Why should these nations defer the pursuit of economic opportunities for the sake of protecting largely obscure, often unknown numbers and kinds of mainly invertebrate species? Calls for sacrifices of this scope and nature tend to be especially unconvincing when the benefits derived from biological diversity are abstractly formulated, projected into an uncertain future, and dispersed to humanity in general. Developing nations are often reluctant to bear the burden of this denial in the face of agonizing pressures of poverty and overpopulation. The creation of "paper" parks on marginal lands should not be regarded as much cause for optimism.

The primary tension in developing nations is reconciling protectionist objectives—endangered species preservation, biological diversity, ecosystem maintenance—with the utilitarian values of often impoverished, unhealthy, subsistence-oriented populations. It may therefore be necessary to link any species conservation strategy with socioeconomic programs that address such basic problems as: low per capita incomes, rural underdevelopment, small-scale subsistence agriculture, slow economic growth, nutritional deficiencies, poor health, high rates of population growth, high rates of unemployment, large dependence on primary product exports (e.g., minerals), dependence and vulnerability in international trade relations, dependence on firewood for cooking and heat energy, and grossly inequitable distributions of national income.[21] Additionally, species preservation efforts in the less developed countries may have to be designed in ways that directly assist indigenous populations rather than primarily serving the needs of foreign tourists, scientists from the developed nations, or a central government.

Several of these problems were recently described in Kenya, one of the few developing nations to derive major financial benefits from land and species preservation. Lusigi concluded that the very concept of species and nature preservation "as presently conceived is an alien and unacceptable idea to the African population."[22] Most Kenyans tend to perceive the conservation of land and wildlife as a choice of animals and plants over people. Creating parks and preserving species as a means of promoting tourism (and, thus, obtaining hard currency) is often viewed as taking land from local populations for the benefit of foreigners. From this perspective, protected areas and species may be regarded as residues of the colonial era.

Unfortunately, it is true that the establishment of nature preserves in developing countries has sometimes resulted in local peoples being displaced from resources and lands of traditional, cultural, and practical value, with income derived from the protected lands being distributed to urban administrators and central government officials. The alienating tendency of such arrangements is suggested by Western and Henry who note:

> Even when they do not take up land that could be more profitably used for other purposes, most parks in developing countries form enclaves in which the economic disparity in earnings between the park and its environs is enormous. Under these circumstances, the park is seen as a resort for foreign clientele, which has little relevance or benefit for the local populace, even if it does for the national government. This inevitably alienates the parks from the very populations on which they are most dependent for survival.[23]

Several strategies have been suggested for mitigating these adverse impacts and enhancing support for species and land preservation in developing countries. One possibility is some form of financial compensation for socioeconomic losses associated with efforts to protect endangered species. International aid agencies, for example, could administer incentive funds for protecting areas and species of significant ecological value. Myers additionally suggests levies or taxes on transnational corporations deriving direct monetary benefits from the resources of relatively pristine ecosystems—for example, pharmaceutical, agribusiness, forestry, and mineral-related companies.[24] Finally, a share of the profits obtained from wildlife-related tourism could be distributed to local peoples displaced or adversely afffected by the creation of preserves.

The major challenge in less developed countries remains the need to incorporate the socioeconomic aspirations and utilitarian values of the general population into the establishment and management of preservation programs. It may be necessary to include rather than exclude local peoples from protected areas, but somehow within the carrying capacity and renewable resource limits of the land. This approach would emphasize the maintenance and perhaps enhancement of living standards among indigenous populations as a concurrent objective of species and habitat protection. The primary virtue of this conservation strategy, according to Lusigi, is that:

> Instead of being isolated islands, the [protected areas and species] would be integral parts of the land use of the whole area, contributing to the social and economic development of the area where they are located. . . . Few developing nations can afford the luxury of [land or species protection] for purely esthetic or philosophical reasons. It is essential to consider local economics and local traditions in addition to ecological factors.[25]

Some degree of subsistence use (e.g., hunting, livestock grazing) and even economic development might be permitted within the boundaries of protected areas, but it would have to be strictly monitored and subjected to enforceable regulations.

The "biosphere reserve" concept put forward by the International Union for the Conservation of Nature and Natural Resources embraces this kind of conservation strategy. It explicitly emphasizes the necessity of heeding the socioeconomic needs and environmental perceptions of local people in the creation of preserves: "Social and economic activities of local populations comprise a significant management imput. . . . the emphasis of the program is on the relation between man and nature. To be successful, it must preserve areas of undisturbed nature as genetic reservoirs . . . [but] it must equally include man and his works."[26]

The potential utility of this strategy was recently suggested by the experience of Mexico's Mapimi Biosphere Reserve. This approach was chosen after the administrators noted the following limitations of conventional Western land protection schemes:

> In rich countries . . . a traditional scheme for [species and land protection] can be effective. . . . The situation in many . . . developing countries is different. Demographic pressures force landless peasants, whose only chance for survival is subsistence agriculture, to occupy protected areas illegally. . . . Hunting is not efficiently controlled . . . trees are cut . . . frequently tourism is not well organized. . . . It would be easy to blame these irregularities on the administrative authorities concerned. . . . The fundamental problem, [however,] is that the concept on which these parks are based does not address the problems of conservation in developing countries in a full and flexible way.[27]

The creation of the Mapimi Biosphere Reserve involved extensive contacts with a variety of social and political groups at the local and federal level. In addition, traditional economic uses and resource values of the areas were considered in the preserve plan, with the raising of the local population's per capita income identified as a primary objective of land protection. Thus, Halffter notes: "One of the fundamental objectives of the two Durango biosphere reserves was to raise the economic and social level of the people living both in the reserve area proper and in the surrounding area."[28] Improvement of livestock production and biological studies of indigenous plants and animals as potential sources of food and income were deemed major considerations of the biosphere research program. These goals were facilitated by a land and species protection strategy which established a core area barred to most human use, surrounded by buffer zones open to more intensive human activity.

The Mapimi experience suggests that a major virtue of the biosphere reserve concept is its potential for reconciling conservation and development goals within the overall objective of land and species protection. By explicitly emphasizing the importance of local values and socioeconomic needs, the biosphere reserve strategy attempts to make land protection a socially familiar rather than alien concept. As Lusigi points out:

> Great emphasis is placed on the co-operation and participation of the local population. Scientific research is not confined to the ecology of the flora and fauna, but also concerns the rational use of natural resources in connection with the economy of the surrounding region.[29]

The success of land and species protection in developing countries will depend, as well, on the emotional and ethical convictions of the general population. Local support should be enhanced by active involvement in the planning and management of species and habitat preserves. In addition, it would be worthwhile to use moral persuasion and psychological reward as important elements in planning. One possibility for increasing awareness and concern is to emphasize a developing country's biological resources as a basis for national pride and cultural distinctiveness. In a world tending toward an increasing homogenization of values and customs, and a consequent erosion of national identity, the uniqueness of indigenous fauna and flora can be cited as one important distinguishing characteristic of a nation. The potential of this approach is suggested by Norman Myers's remark: "By virtue of their diversity of species, many developing nations of the tropics are the equivalent of 'biological millionaires' as compared with the ecologically impoverished nations of the rest of the world."[30] Employing this approach in Madagascar, local support for species protection was strengthened because the people became proudly aware that, despite their country's political and socioeconomic backwardness, the island was among the world's most outstanding biological areas.[31]

An additional point to consider is the nature of the economic, especially trade and monetary, relationship of the developing countries with the industrial powers. It would appear somewhat counterproductive, if not hypocritical, for Western nations to advocate environmental protection, if their economic demands on the developing countries encourage excessive resource exploitation and degradation. In this regard, Clement insightfully notes:

> We blame what seem to us insensitive attitudes toward forest destruction by the [peoples of developing nations], or we point to overpopulation as a basic cause. We must also learn to recognize that today's forces of destruction are a byproduct of our own economic demands. It was partly to facilitate mining activities by foreign capital that Brazil's transAmerica highway network was built; and it is partly to satisfy America's hamburger culture that cattlemen have displaced . . . [large areas] of central America.[32]

Developed nations need to promote policies and practices among their business and bilateral lending agencies that minimize destruction of wildlands and pollution in the developing nations. We need to consider whether the various international aid agencies, multinational corporations, and commercial banking institutions have shown sufficient concern for the ecological consequences of their actions. Unfortunately, studies of the environmental practices of diverse multinational corporations and of bilateral lending policies among a cross-section of Western countries have found only marginal attempts to consider ecological factors in a variety of resource-related projects in developing countries.[33] When environmental factors were assessed, the focus was almost exclusively on the impacts of air, water, and toxic waste pollution on human well-being, and rarely extended to consideration of the nonhuman inhabitants of affected areas. For the most part, environmental considerations were ancillary rather than integral components in the planning and evaluation of economic projects in developing countries.

The relationship of economic markets for products taken from the wild to excessive exploitation of wildlife in the underdeveloped countries also needs to be considered. As a result of increasing technological access to remote areas, and swelling consumer demands in the industrial countries, the international trade in wildlife products is now estimated to be in excess of one billion dollars per year.[34] The taking of wildlife for a variety of skins, shells, bones, furs, and other products is currently cited as the primary causal factor in 37 percent of all vertebrate species endangerments.[35]

Consumer demand in the industrially developed nations largely accounts for the incentive to market these wildlife products. While an ethic of environmental appreciation has increasingly manifested itself in many economically advanced countries, an ironic consequence of this new wildlife awareness may be an expanded interest in objects associated with the natural world. For example, a shift in demand for clothing made from domestically raised furbearers to fur products derived from long-haired wildlife species may represent one consequence of this attitude change. Ths Convention in International Trade in Endangered Species has helped to discourage the marketing of products derived from rare and threatened animals, although history has repeatedly demonstrated that the existence of strong economic rewards typically provides sufficient incentive to circumvent regulatory controls.[36] It may thus be necessary to consider the basis for these consumer demands in the developed countries, in addition to the policies and practices of developing countries that supply these markets. According to Clement: "The new traffic in animals and their parts we need to confront and regulate is a by-product of the jet age and the mass-consumption society. It is the result of uneducated affluence."[37]

IV. CONCLUSION

Several socioeconomic and perceptual factors have been identified and discussed as requiring consideration in any attempt to develop effective solutions to the species extinction crisis. As a way of organizing these factors more systematically, we may separate them into three categories: socio-psychological, economic, and administrative-political. Socio-psychological factors include: people's attitudes toward rare and endangered wildlife; basic land-use values; attitudes toward governmental regulation and control; and perceptions of widely varying life forms. Economic factors that have been identified or alluded to include: economic dependence on natural resource exploitation; the distribution of private and public property; the relative importance of agricultural and mineral extraction; the economic contribution of species; the amenity enjoyment of nature; and economic dependence on publicly owned lands. Finally, political-administrative variables include: the promulgation and effectiveness of land-use and wildlife protection regulatory standards; the relation of political power to resource extraction; zoning and land-use laws; bureaucratic role distribution and decision-making with respect to wildlife and endangered species management; and the historical context of environmental conflict resolution.

These various socio-psychological, economic, and political-administrative factors will need more thorough description and analysis if an adequate model of social factors in preservation issues is to be developed. The preliminary discussion offered here only suggests possible strategies for easing the increasingly serious problem of endangered species. Some implications of the discussion are listed as follows.

1. We need to articulate and specify more precisely all values people associate with wildlife and the benefits derived from its preservation.

2. Whenever possible, these values and benefits should be quantitatively and empirically measured.

3. These values should be consistently considered when choices between species and habitat preservation and socioeconomic development are being made.

4. We need to recognize and understand that value attributions vary considerably depending on the species, particularly when comparing vertebrate and invertebrate animals. This variation should be given systematic consideration in assigning species preservation priorities and in designing public education and awareness programs. In certain situations, this assessment might result in a greater emphasis on protecting particularly valued "higher" life forms. Conversely, it might require extensive public education efforts to protect species with especially negative images or that are phylogenetically remote from human beings.

5. As most of the planet's endangered species are invertebrates, we need to develop a compelling rationale for their preservation. Animals with backbones are, for the most part, valued for ecological, scientific, aesthetic, and utilitarian reasons. Only very few people recognize, how-

ever, the ecological and scientific worth of invertebrates. Invertebrate species preservation efforts should therefore emphasize the aesthetic and especially utilitarian contributions of these life forms to human welfare and economic well-being.

6. Most endangered invertebrates exist in the tropical moist forests of the less developed countries. The needs and perceptions of people in these countries have to be given special consideration in global species preservation efforts.

7. The dominant wildlife-related values of people in the less developed countries are sacral/mystical and utilitarian. Wildlife protection programs in these countries should emphasize these rather than Western-oriented values.

8. A strong utilitarian perspective in the less developed countries suggests that endangered species and land protection programs should be incorporated into socioeconomic development schemes.

9. Economic ventures that foster wide-scale tropical habitat destruction and consequent extinctions, and the conditions that make such ventures possible, should be scrutinized. Among the areas requiring attention are direct exploitation of wildlife products to serve consumer markets in the industrially developed economies, the environmental policies of resource-based multinational corporations, and the financial lending practices of commercial banking and multilateral aid institutions.

While the emphasis of this chapter may have been somewhat pragmatic, the long-term solution to the global extinction crisis will probably depend more on our developing a closer personal and spiritual sense of relatedness to the nonhuman world. The evolution of a gentler ethic and a feeling of empathy and kinship for all living things may be the ultimate prerequisites of any successful species preservation effort. If this change occurred, it would probably result in an awareness that the elimination of any species also represents a diminution in the quality of human life. Extinction would be regarded not just as a reduction in biological options for coping with an uncertain future but, more important, as a closing off of the aesthetic, cultural, and spiritual opportunities that humans crave in their quest to make life more meaningful. Perhaps Leopold best captures the ambivalence of any irreversible exchange which sacrifices the future for the present, in his commemorative remarks on the passing of "Martha," the last of the passenger pigeons:

> We grieve because no living man will see again the onrushing phalanx of victorious birds sweeping a path for Spring across the March skies, chasing the defeated winter from all the woods and prairies. . . . Our grandfathers, who saw the glory of the fluttering hosts, were

less well-housed, well-fed, well-clothed than we are. The strivings by which they bettered our lot are also those which deprived us of pigeons. Perhaps we now grieve because we are not sure, in our hearts, that we have gained by the exchange. . . . The truth is our grandfathers, who did the actual killing, were our agents. They were our agents in the sense they shared the conviction, which we have only now begun to doubt, that it is more important to multiply people and comforts than to cherish the beauty of the land in which they live.[38]

NOTES

1. Andrew Berger, *Hawaiian Birdlife* (Honolulu: University Press of Hawaii, 1972); and Winston Banko, "Historical Synthesis of Recent Endemic Hawaiian Birds," Cooperative National Parks Resources Study Unit, University of Hawaii, Manoa, 1979.
2. Norman Myers, *The Sinking Ark* (New York: Pergamon Press, 1979); and Gerald O. Barney, *The Global 2000 Report to the President* (Washington, D.C.: U.S. Government Printing Office, No. 0-256-752, 1980).
3. Stephen R. Kellert, "Contemporary Values of Wildlife in American Society," in *Wildlife Values*, edited by W. W. Shaw and E. H. Zube (Fort Collins, Colorado: USDA Forest Service Center for Assessment of Noncommodity Natural Resource Values, 1980), pp. 31-61; R. T. King, "The Future of Wildlife in Forest Use," in *Transactions of the North American Wildlife Conference* 12 (1947): 454-66; and Holmes Rolston III, "Values in Nature," *Environmental Ethics* 3 (1981): 113-28.
4. David Lime, "Wildlife Is for Nonhunters, Too," *Journal of Forestry* 74 (1976): 600-604.
5. Michael Fox, *Returning to Eden: Animal Rights and Human Responsibility* (New York: Viking Press, 1980).
6. Myers, *Sinking Ark*; and Paul Ehrlich and Anne Ehrlich, *Extinction: The Causes and Consequences of the Disappearance of Species* (New York: Random House, 1981).
7. General Accounting Office, *The Tennessee Valley Authority's Tellico Dam Project—Cost Alternatives and Benefits* (Washington, D.C.: Report No. EMD-77-58, 1977); Endangered Species Staff Report, *Tellico Dam and Reservoir* (Washington, D.C.: U.S. Fish and Wildlife Service, Office of Endangered Species, 1979).
8. Stephen R. Kellert, *Public Attitudes Toward Critical Wildlife and Natural Habitat Issues* (Washington, D.C.: U.S. Government Printing Office, Superintendent of Documents, STOP: S.S.M.C., No. 024-010-623-4, 1980).
9. Stephen R. Kellert, "Public Attitudes, Knowledge and Behaviors Toward Wildlife and Natural Habitats," in *Transactions of the North American Wildlife and Natural Resources Conference* 45, 1980: 111-23; and Robert

Cameron Mitchell, "The Public Speaks Again: A New Enviromental Survey," *Resources* 60 (1978): 1-10.

10. Aldo Leopold, *A Sand County Almanac* (New York: Oxford University Press, 1949; reprinted 1978).

11. See Bryan G. Norton, this volume.

12. Thomas Lovejoy, personal communication.

13. Thomas Allen, *Vanishing Wildlife* (Washington, D.C.: National Geographic Society, 1974).

14. Bryan G. Norton, "Environmental Ethics and Nonhuman Rights," *Environmental Ethics* 4 (1982): 18-36.

15. John N. Martin, "The Concept of the Irreplaceable," *Environmental Ethics* 1 (1982): 40.

16. Myers, *Sinking Ark*; and Ehrlich and Ehrlich, *Extinction*; Edward O. Wilson and Peter Raven, Testimony before the Senate Subcommittee on Environmental Pollution, Committee on Environment and Public Works, United States Senate, Ninety-Seventh Congress, first session, December 8 and 10, 1981. Printed in *Endangered Species Act Oversight* (Washington, D.C.: U.S. Government Printing Office, 1982), pp. 268-302, 367-79.

17. Raven, Congressional testimony.

18. Wilson, Congressional testimony.

19. Paul Shepard, Jr., *Thinking Animals* (New York: Viking Press, 1978).

20. Gordon M. Burghardt and Harold A. Herzog, Jr., "Beyond Conspecifics: Is Brer Rabbit Our Brother," *Bioscience* 30 (1980): 763-68.

21. Michael P. Todero, *Economic Development in the Third World* (New York: Longman, 1980).

22. Walter J. Lusigi, "New Approaches to Wildlife Conservation in Kenya," *Ambio* 10 (1981): 88.

23. David Western and Wesley Henry, "Economics and Conservation in Third World National Parks," *Bioscience* 29 (1979): 414-18.

24. Myers, *Sinking Ark*, p. 99.

25. Lusigi, "New Approaches," p. 90.

26. International Union for the Conservation of Nature and Natural Resources, *The Biosphere Reserve and Its Relationship to Other Protected Areas*, UNESCO, 1979, p. 18.

27. Gonzolo Halffter, "The Mapimi Biosphere Reserve: Local Participation in Conservation and Development," *Ambio* 10 (1981): 93-96.

28. Ibid., p. 96.

29. Lusigi, "New Approaches," p. 91.

30. Myers, *Sinking Ark*, p. 99.

31. Alison Richard, personal communication.

32. Roland Clement, "Culture and Species Endangerment," in *Proceedings of the Endangered Species: A Symposium, Great Basin Naturalist Memoirs* 3 (1979): 12.

33. See Thomas N. Gladwin, *Environment Planning and the Multinational Corporation* (Greenwich, Conn.: JAI Press, 1977) on the environmental practices

of multinational corporations; and Brian Johnson and Robert Blake, *The Environment and Bilateral Aid* (Washington, D.C.: International Institute for Environment and Development, 1979), on bilateral lending policies.

34. F. W. King, "The Wildlife Trade," in *Wildlife and America*, edited by H. P. Brokaw (Washington, D.C.: Council on Environmental Quality, 1978), pp. 253-71.

35. International Union for Conservation of Nature and Natural Resources, *World Conservation Strategy* (New York: Unipub, 1980).

36. Peter Matthiessen, *Wildlife in America* (New York: Viking Press, 1959), p. 107.

37. Clement, "Culture and Species Endangerment," p. 12.

38. Aldo Leopold, "A Passing in Cincinnati" (U.S. Government Printing Office Stock 024-000-008244-0, 1964), p. 15.

PART II

Values and Objectives

Introduction to Part II

The chapters in Part I of this book attempt to delineate the endangered species problem. The assessment of an event or trend as a problem implies that the current situation is not ideal and some values are not fully satisfied. That there is broad agreement that diminishing biological diversity is a problem, however, does not imply that there is a single value served by that diversity.

One can distinguish anthropocentric and intrinsic values. Anthropocentrists believe that all values are ultimately human values, that non-human species and other natural objects have value only instrumentally, for the fulfillment of human needs. This position denies that other species are the locus of intrinsic value. Within this realm, one can further distinguish utilitarian values, those that fulfill material needs, from those that elevate human consciousness—aesthetic values, the value of knowledge, and so forth. It is possible, however, to maintain that natural objects, including species, have intrinsic value, providing checks and limits upon human use of them. Both kinds of values are discussed in Part II.

Alan Randall, writing from an economist's viewpoint, insists that it is possible and useful to treat all these values as preferences of human choosers. Values, whether consumptive, aesthetic, or altruistic, can be quantified in the common terms provided by economic techniques.

The advantages of quantification are considerable: the costs and benefits of actions designed to protect species can be thereby, in principle, compared. Randall argues that, if the fullest range of human values is included in the comparative analysis, the preservation option will fare well in many decision situations.

Bryan Norton argues that a wide range of easily underestimated utilitarian values can be derived from other species and the ecosystems they compose. He believes that the tendency to focus the utilitarian question on the values of preserving individual species necessarily undervalues them by failing to take into account the ways in which each species contributes to the development and functioning of ecosystems. Further, he believes that attempts to place dollar values on ecosystems are doomed to failure and thus he advocates a strong principle of preservation—that all species should be saved, provided the costs are tolerable.

J. Baird Callicott considers the difficult notion that other species have rights that constrain our treatment of them, and concludes that references to rights in this context are best interpreted as a means to attribute intrinsic value to nonhuman species. He examines several axiological theories that would support such value. The strongest defense of attributing intrinsic value to nonhuman species, he thinks, is provided by David Hume's theory of values based on human moral sentiments, combined with an expansion of Darwinian insights concerning the origins of these sentiments. Callicott's theory, then, sees nonhuman species as a locus of intrinsic value without insisting that this value is entirely independent of human sources.

Elliott Sober rejects appeals to the intrinsic value of nonhumans as both difficult to defend and unnecessary. After surveying a number of standard arguments offered to support species preservation, he concludes that the concern of environmentalists is best understood as aesthetic in nature. The damage done when a species becomes extinct is analogous to the damage done when a great work of art is destroyed.

Donald Regan develops a novel approach to the value of nonhuman species, arguing that, while species themselves are not intrinsically valuable because all value depends upon human valuers, the complex event formed by a natural phenomenon such as a species, a human being's knowledge of it, and that human being's pleasure resulting from such knowledge, does have intrinsic value. Destruction of species decreases the number and type of such complexes possible and, consequently, decreases the potential intrinsic value in the world.

All the authors discussing values and objectives in relation to species preservation shy away from value wholly independent of human sources. This no doubt results at least in part from the difficulty of defending values independent of conscious valuers. But at the same time, all authors show concern that values not be limited to too narrow a range. The possibility raised by Callicott and Regan that intrinsic value, while not independent of human valuing, may be attributed to, or augmented by, nonhuman species illustrates this concern. It indicates that the dichotomy between anthropocentrism and other viewpoints is neither as sharp nor as important as is sometimes assumed. It seems clear that protection of other species can be justified by appeal to complexes of reasons deriving from utilitarian considerations narrowly construed plus a broader range of values not easily characterized in terms of immediate human needs. All authors advocate inclusion of values which are in some sense altruistic or aesthetic, rather than purely prudential.

4

Human Preferences, Economics, and the Preservation of Species

ALAN RANDALL

I. Preservation of Species as a Resource Allocation Problem

The earth's biota may be viewed as a resource or a complex group of resources. This view carries two immediate implications. First, biotic resources are instruments for human satisfaction and second, they are scarce. Scarcity means they are both valued and limited and that they can be increased only at the cost of forgoing something else that is valued.

It has been customary to refer to biotic resources as renewable. More recently, the term "renewable but destructible" has come into favor, reflecting the realization that there is nothing assured about renewal. The "renewable" side of the coin refers to an ecosystem's capacity to evade the entropy law and, by capturing energy from the sun and recycling water and nutrients, establish a continuity in which each generation "lays the seeds" for the next. The "destructible" side of the coin recognizes that continuity requires favorable conditions and can be interrupted by a variety of problems, many of which spring from an excess of human attention (e.g., excessive harvesting or application of pesticides) or inattention (e.g., inadvertent destruction of habitat).

The preservation of species is one part of the larger problem of allocating biotic resources. It can be seen as a threshold problem. Presumably, there is some threshold population size below which a species tends inexorably toward extinction, but above which species survival is possible. Since species survival is a precondition for all use of the species as a resource, the preservation problem in principle precedes all other biotic resource issues.

Biotic resources are not the only resources: we have many other kinds which may serve as instruments for human satisfactions. Further, biotic resources may serve in various ways. The potential for conflict over the allocation of all available resources is obvious. Ultimately, the question is how the human inhabitants of a complex world may best satisfy their desires. Biotic resource allocation issues, including those associated with the preservation of species, must be resolved within this broader context.

In this chapter, I attempt to explore what can be learned by treating preservation issues as a problem in resource allocation. In so doing, I recognize that other perspectives are legitimate. For example, one may treat nonhuman life forms as in some sense co-equal with humans. That is, each life form may be seen as possessing rights, and/or having "a good of its own." These viewpoints have been evaluated elsewhere, as have perspectives based on ethical duties to species arising from human purposes that are not merely instrumental.[1] I choose not to address these alternative or additional perspectives, concentrating instead on the implications of the resource allocation approach.

This is no easy task, because species preservation issues by their very nature must be resolved by collective human decisions. The "invisible hand," which works fairly efficiently to guide individual actions in many spheres, cannot resolve fundamentally collective problems. That does not mean economics is silent on preservation issues. But it does mean that the economic arguments are necessarily complex and not entirely conclusive. It is essential to work through these arguments, as much to understand their limits as to appreciate their conclusions.

The discussion proceeds as follows. In Section II, the economic theory of individual choice is introduced, to develop the concept of value and identify its normative implications. The kinds of economic value that pertain to biological species are discussed. Since economic value is based on human preferences, some conjectures are offered about the kinds of preferences that might come into play. Section III is devoted to normative economic theories of collective choice, to determine the limits of economic logic in providing prescriptions for public choice. In economics, as in other branches of philosophy, no normative theory of collective choice can be proven right, or superior to all others, and none commands universal assent. An interesting finding is that the benefit cost (BC) criterion can be derived from one branch of utilitarian philosophy. That means that proposals to base public decisions on the outcome of BC analyses can be criticized on the grounds that the BC criterion emerges from an unacceptable normative theory of government, but not on the grounds that BC analyses (being pertinent only to the market sphere) are strictly irrelevant to public choices. Abandoning the search for normative truth, in Section IV we consider the kinds of economic information that can be brought to bear on public decisions about the preservation of biological diversity, and how the public decision process might use that information. Potential roles for BC information are assessed. In addition, a perhaps lesser known approach, based on the concept of a safe minimum standard of conservation, is introduced and its implications discussed. In Section V, an attempt is made to develop a system of preservation priorities. Considerable progress is made (I believe) in ordering priorities, but very little can be said about the overall cut-off point for preservation efforts. Finally, in Section VI, the argument is summarized in terms of some fifteen concluding statements.

II. THE ECONOMICS OF INDIVIDUAL CHOICE

The discipline of economics is on relatively firm ground when it considers individual choice in an environment of scarcity. Scarcity implies limits, or constraints, and constrained choice is guided by preferences.

Human Preferences

Nothing is more basic, in a mainstream neoclassical economic model, than preferences. The individual is presumed to be able to identify alternatives, rank them in a complete and transitive manner, and choose among them on the basis of these preference rankings. Mainstream economics ascribes considerable normative import to individual preferences. A statement frequently encountered may be paraphrased: the individual is the best judge of what is good for him/herself; therefore whatever a person wants is presumed to be good for that person. This position is based on two quite separate premises. First, it is argued (or sometimes merely assumed) that people come by their preferences in a serious manner, reflecting cultural traditions, individual learning and experience, and a good deal of introspection.[2] That is, preferences are not whimsical or capricious.[3] Second, there is the strongly normative premise that no one else is better qualified than the individual to know what is good for him or her.

Some discussions contrast "changeable preferences" with attempted answers to the fundamental question of ethics, "What shall I do with my life?"[4] This dichotomy is presumably intended to emphasize that preferences have little moral force. But surely the dichotomy is overdrawn. It seems to me that the individual's structure of preferences represents his/her current and tentative best answer to the "What shall I do with my life?" question. Preferences are expressed not merely across material goods but also across lifestyles and concepts of deeper purpose: not just for foodstuffs, the latest fads in attire, and the like, but also concerning the well-being of one's contemporaries and future generations and alternative conceptions of human relationships.

This is not to place individual preferences on a par with ethical principles; at best, they represent individual and tentative answers to the fundamental question, not a general and reasoned resolution thereof. Nevertheless, and especially while general ethical principles remain elusive, individual preferences have some moral force.

Constrained Choice and Economic Value

The notion of scarcity is ultimately brought to bear on the individual via constraints on the endowments at his or her command. Thus, choice is constrained choice. The choice bundle (the totality of things chosen) is determined by the individual's preference rankings, endowments, and the costs (in term of endowments) of the various alternatives which comprise the opportunity set.

If myriad individuals make choices in an environment where efficient markets accurately signal the costs of alternatives, the prices that emerge reflect economic values.

Prescriptive economic analyses usually treat the results of constrained choice as having some normative significance. Since preference rankings are an important determinant of choice, the limits to their normative significance, discussed above, apply. However, there are additional limits to the normative significance of the results of choice in markets. First, individual choice in markets is constrained by endowments: by what kinds of endowments the market counts, the weight it places on each kind, and the individual's allotment of them. Those who perceive massive injustice in the way endowments are distributed among individuals and valued in markets tend to discount the normative significance of constrained choice for that reason.

Second, for some important categories of things people care about, markets are limited and there is only a weak basis for faith in their performance. Nonrivalry, nonexclusiveness, and decisions involving very long time horizons, for example, introduce major impediments to the functioning of markets. Unfortunately, these difficulties are endemic to preservation issues.

(1) As opposed to ordinary (rival) goods, where for a given level of output more for one person necessarily means less for others, the preservation of species is a *nonrival* good. Once preservation has been provided, there is no meaningful sense in which it must be divided up among individual "consumers" of preservation. Once provided, it is effectively provided for all.

(2) The existence of a species is usually a *nonexclusive* good. That is,

there is no meaningful way that some people (e.g., those who make no contribution) can be excluded from enjoying the satisfaction of knowing the species survives. However, some uses to which an organism may be put (e.g., in the plant breeding, pharmaceutical, and genetic engineering industries) are amenable to exclusive arrangements.

(3) Since there seems no reasonable prospect of a technology for precisely recreating lost life forms, a brief lapse of vigilance and commitment to preservation would have enormously long consequences. An irreversible loss would be inflicted on every subsequent human generation. Markets, and market-oriented economic theories, tend to treat all intertemporal allocation problems as investment problems. However, it is simply not clear that investment logic is adequate to handle issues such as preservation, where *time horizons are very long* and losses may be *irreversible*.

These various "market failures"[5] do not directly threaten the economic concept of values. They do, however, generate serious impediments to market revelation of economic values relating to preservation issues.

More sophisticated analysis of constrained choice recognizes that choice in an environment of scarcity inevitably leads to conflict among individuals, and that markets are merely one kind of conflict resolution institution. Others include elective politics, the legislature, other branches of government, and a bewildering variety of special-purpose institutions. Individuals will seek satisfaction through these various institutions, allocating their effort and endowments across arenas on the basis of their expected productivity in each. Coalitions, ephemeral and more permanent, may be formed, as individuals find the gains from membership exceed the cost. Individuals and coalitions will seek to maximize their satisfaction through existing institutions and by investing in institution-changing behaviors. If one's endowments are more effective in one kind of institutional arena than another, one might rationally work toward increasing the range of conflicts which come under the purview of the first kind.[6]

Economic Values Associated with Preservation

The economic value to an individual of an increment in any good or amenity is the maximum amount of money he/she is willing to pay (WTP) for it. The value of a decrement is the minimum amount of compensation which would make the individual willing to accept (WTA) that decrement.[7] These value measures are conceptually valid, whether or not adequate markets exist in the good or amenity of interest. However, efficient markets, where they exist, are a convenient source of observations which can readily provide approximations of WTP and WTA.

The concept of cost is directly derived from that of value. In an environment of scarcity and constrained choice, selection of one alternative comes at the expense of forgoing other opportunities. The cost of alternative A is, then, the net value of all the things which must be forgone if A is chosen. Economists call this *opportunity cost*.

The import of this discussion is that whatever is wanted has some economic value (perhaps less than its cost, but nevertheless some value). Biotic resources may be valued for a variety of uses and purposes. These categories include:

Use value. This refers to economic value derived from using the ecosystem in any way: as a source of raw materials, medicinal products, aesthetic satisfaction, personally experienced and vicarious adventure, etc. Use value may be derived from present and expected future use.

Option value. When risk attends the demand for use and/or the supply of ecosystem services, expected values from future use must be modified to include risk discounts and premia.[8] For example, consider a risk-averse individual who is sure to demand future use but unsure that the resource will be available then. This person might rationally pay something beyond future use value to secure an option guaranteeing the resource will be available for later use.

Quasi-option value. If more information is expected to be revealed with the passage of time, delay in deciding the disposition of some habitat may result in a better decision. Quasi-option value can be considered the value of preserving options, given the expectation of growth in knowledge.[9] It is reasonable to expect that, for example, the list of species known to provide pharmaceutically useful chemical compounds will grow with the accumulation of research results over time. For any species there is some positive, if unknown, probability that new uses generating positive value will eventually be discovered. With extinction, however, that probability drops to zero, as does the expected value of the resulting benefits.

Existence value. This is the value enjoyed from just knowing that (something) exists. Since the concept of existence value is susceptible to misinterpretation, a few clarifying comments may be helpful.[10]

(i) Existence value for a resource must (in order to distinguish it from use value) be derived without combining it with any other good or service in order to produce any activity other than "just knowing it exists."

(ii) Existence values are not confined to unique natural phenomena threatened with irreversible damage.[11] In principle, anything may have existence value. However, value at the margin depends on demand and supply. It is likely that the existence value, in total, for cattle exceeds that for condors. However, the supply of cattle is very large while the supply of condors is quite small. Thus, the loss in existence value from the destruction of a few condors likely far exceeds that from the loss of a few cattle.

(iii) Since existence values are independent of current use, expected future use, and the avoidance of risks related to future use, they must be derived from some form of altruism. Three relevant kinds of altruistic motives come to mind: *philanthropic*, in which the resource is valued because one's contemporaries may want to use it; *bequest*, in which the source of value is that future generations may want to use the resource; and *intrinsic*, in which the individual human cares about the well-being of nonhuman components of the ecosystem. This last-mentioned kind of altruistic motivation is consistent with both my earlier definition of species as resources (i.e., instruments for human satisfaction) and Regan's denial of the notion that a nonhuman entity may have "a good of its own."[12] Intrinsic altruism is consistent both with Regan's approach since the human remains the valuer, and with mine, since caring is extended because it gives the human satisfaction to do so. Such caring may, of course, be limited and selective, and since it is extended unilaterally it may be similarly withdrawn. The recognition of intrinsic altruism in no way endorses concepts such as "species rights."

The Information Base for Valuation

Biological species are valued by people because they contribute directly to human satisfaction, or because they play an important role in the life support systems of other species that do so. Valuations depend crucially on information and knowledge. Individuals place no value on resources of whose existence or usefulness they are entirely unaware. As information (of varying degrees of reliability) is acquired, valuations may be quite volatile. Eventually, as knowledge of consequences becomes more complete, valuations tend to stabilize and to reflect more accurately the individual's underlying preferences and endowments.

The information base for valuation of species is very weak. Perhaps seventy percent of the earth's species are as yet uncataloged. For many species that are known, we have little appreciation of their roles in life support systems or the services they might provide directly to humans. The multiplicity of direct and indirect services provided by those species that have been carefully studied suggests that the biota contains much of value that we do not yet know about. Finally, if we seek to evaluate not so much species per se as policies to protect them, we need information about crucial relationships within the system. What does it take to assure the survival of species X? How large must its range be? How can we be sure not to tip the delicate balance inadvertently in favor of its predators or competitors? If species X is saved, what else might inadvertently be saved, or lost, as a result? Can the ecosystem survive the loss of some species? If so, how many can be lost before total collapse? Are all species equally crucial to the survival of the ecosystem or, if the ecosystem's survival is the goal, are some species more expendable than others? I do not mean to imply that nothing is known about these things.[13] However, there is much that remains unknown, and it is reasonable to expect valuations to change as more information becomes available.

Limits on information provide yet further restrictions on the normative significance that can be placed on individual choices and valuations regarding preservation of species, since the full range of consequences of acts that may threaten a particular species can seldom be identified even in the broadest terms. However, the solution is unlikely to be found in delegating these issues to the professional experts. Confusion among the citizenry often reflects incomplete and dissonant signals from the specialists, while lay individuals (at least in modern societies) fairly rapidly reflect expert opinion when it is confidently held, convincing, and near-unanimous. Unfortunately, in matters concerning preservation of species, it is easy to bemoan the ignorance of the laity, but harder to claim that broad-based and reliable knowledge exists elsewhere. The solution to the "ignorance" problem, it seems to me, will require increased support for the esoteric enquiries of specialists and efforts to communicate their findings more rapidly to the general public.

Preservation of Species: What Do People Value?

The biotic resources model views species as instruments for human satisfaction. However, as was indicated by the discussion of the categories of value associated with preservation, the model places few a priori restrictions on the human motivations considered. A wide range of motivations may find expression in values assigned to species. Nevertheless, it is useful to speculate about which preferences are most prevalent. These are the preferences which will, other things being equal, be influential in economic valuation.

First, what do we mean by preservation? The ecosystem is dynamic, constantly changing and evolving. Its rate of change, however, is by no means constant. The fossil record reveals periods of convulsive shock, as well as periods of slow evolutionary change. Major extinctions have occurred in past eras, e.g., the various ice ages and periods following the introduction of successful predators (including humans) to new territories. Yet subsequent periods of relatively stable conditions have permitted speciation and the increasing diversification of ecosystems.[14] In a sense, it seems that "life goes on," that is, biological life can withstand considerable shocks even though individual life forms are much more fragile.

This suggests two possible meanings for preservation: preservation of currently existing life forms, or species, one-by-one, to preserve the ecosystem *as we know it*; or preservation of the capacity for future evolution, speciation, and diversification. Attempts to implement the former concept would entail eventual conflict with the evolutionary process itself. Humankind would find itself in the position of attempting to impose its preferences (in favor of the ecosystem as we know it) upon a system whose inner dynamic reaches for change.

Consider the following, quite fanciful, scenario. We know that a new ice age is approaching, surely and rapidly, and that many familiar life forms will not survive it. We do not know exactly which ones are doomed, but we know that the survivors will be an impoverished lot and that unfamiliar and unattractive life forms may step into the vacated niches. There is reasonable assurance that most of the human population will survive. Imagine that a technology exists capable of forestalling this approaching ice age, but at very considerable expense. Would we use it? I would expect a substantial consensus in favor of such intervention, including quite a few ecologists and moral philosophers.

What would such a decision imply about human preferences with respect to the ecosystem? Basically, that human preferences are focused on life forms more than on life processes. The coming ice age would not extinguish life processes. If past ice ages are any indication, the resultant ecosystem might eventually exhibit a diversity richer than that which now exists. Nevertheless, we would (I conjecture) choose the familiar life forms—those which we know and care for—over the unknown. We would make a choice, rather than passively accept the verdict of nature. Our preferences favor "us and ours" relative to the unfamiliar and alien.

These conjectures are entirely consistent with the concept of allocatable biotic resources. The familiar are more likely to provide satisfaction, given that techniques and habits permitting people to use and enjoy them are already developed.

Which of the well-known species would we expect to rank prominently in the preference structure of human beings? Those which are domesticated to provide raw materials or amenities for human enterprises would surely rank highly. Some of them enjoy the advantage, in addition to being in some way useful to people, of having established relationships of familiarity and perhaps even mutual trust with them. Undomesticated species with widely recognized raw material or amenity value would probably also be given preference, especially those which have become the stuff of legends and patriotic symbolism, as well as those which are "pretty," "nice," or "cute and cuddly."

Beyond these groups of species, human preferences usually begin to diverge. Those aware of the importance of the gene pool in plant and animal breeding would surely place a high priority on the wild relatives of domestic cultigens. Those who appreciate the contribution of natural organisms in producing pharmaceutically valuable chemicals would want to preserve all species with known similarities to those which have produced useful drugs, at least until they have been screened. Students of nature tend to develop a special appreciation of the objects of their study; these objects are often not merely certain species but entire ecological communities. Ecologists, amateur and professional, operate from the assumption that any living organism must play some role in the scheme of things. (They do frequently choose sides, however, when species come into conflict.) Some people find some familiar and most unfamiliar species in some way threatening; still others simply do not care much, either way, for species whose raw material or amenity value they have yet to recognize.

I have the impression that many of the things the nature lobbies find offensive in current discussions of preservation issues—e.g., "vertebrate chauvinism," concentration on attractive species and those of known usefulness—are actually fairly accurate reflections of many people's current preferences.

Nevertheless, choice and value are not determined by preferences alone, since choices must be made from among the states-of-the-world that are possible. The ultimate instrumental argument for preservation—that every life form must be preserved in order to ensure the survival of the things people do care about and, ultimately, of humanity itself—is based not on preferences but on production possibilities. The argument is not based on the idea that people care for each and every life form. Rather, its basis is that every life form plays some crucial role in the system which supports those life forms we do care about. Thus, this argument provides a rationale—independent of any notions of moral responsibility or obligation—for human concern with the well-being of the vast masses of species which have not impressed themselves on the preference structure.[15]

This rationale for caring about the well-being of species and the integrity of communities and ecosystems is based entirely on fact-statements. It is independent of preferences, save for a preference (believed to be shared by almost all of humanity) for survival. Therefore, it is subject to test in a way that preference-based rationales are not. In principle, it is possible to find out just how strong are the linkages and dependencies among species within communities and ecosystems; whether the loss of any species leads inexorably to systemic collapse; whether some species are more dispensable than others, in terms of the impacts of their loss on the integrity and viability of the system; and what might be the consequences, in terms of the things most crucial to civilized human life, of the loss of particular species or habitats. Answers to these kinds of questions would provide the essential link in determining what we must take care of in order to secure those life forms and ecosystems we do care about.

III. THE NORMATIVE ECONOMICS OF COLLECTIVE CHOICE

Where preferences differ among individuals but circumstances require collective action—for example, in providing important services and amenities that are nonrival and/or nonexclusive—issues arise of collective choice, the relationship of the individual to the group, and the legitimacy of authority. Mainstream economics offers three possible approaches to these issues, each rooted in some variant of eighteenth-century social contract theory, which are discussed below. These different lines of thought in turn suggest different approaches to the preservation of species, habitats, and ecological communities.

Individualism, Voluntary Exchange, and Unanimity

Locke's version of the social contract assigned authority to the individual. Government, deriving its authority from the people, must act only in the public good. Some commentators have seen in Locke's social contract not merely a rationale for the overthrow of governments which have violated the public trust, but strict limits to the authority of any government. That is, individuals are guaranteed certain rights, and a government invading or denying these rights would exceed its rightful authority. Thus, Locke is seen as a founder of philosophical individualism.

Modern individualism, as expressed by, for example, the political economist Buchanan, emphasizes Pareto-safety in economics and in politics.[16] Pareto-safety is the criterion that no change which harms any individual can be considered an improvement. In economic activity, this criterion is satisfied by voluntary exchange among individuals whose expectations are secured by completely specified, enforced, and transferable property rights. Assuming the individual is the best (the only) judge of his own well-being, we can be assured that no one would voluntarily enter a trade which would make him worse off.

In the political sphere, Pareto-safety means that a proposed change must enjoy unanimous consent if it is to be implemented. The idea of the "general will" is explicitly denied. All that matters is the individual, and no individual should be coerced into accepting political change which is not in his or her interest.[17] Recent developments include *incentive-compatible mechanisms*, which in concept permit simultaneous determination of the optimal quantities of public (i.e., nonrival and nonexclusive) goods and optimal individual taxes.[18] In principle, these devices solve the conflict between diverse individual preferences and the necessity of collective action: since each person pays his/her individual WTP (but no more) for goods, each would voluntarily consent to the optimal level of collective provision.

An individualist philosophy would base resolution of conflicts about the preservation of species on voluntary agreement (through trade or unanimous collective choice) based on secure rights assigned at the individual level. Such an arrangement is not inconceivable for one important part of preservation: maintenance of the gene pool and a small number of species in managed environments such as zoos, laboratories, and seed banks. Property rights are assigned, in many regimes, to the results of plant breeding and genetic engineering, in order to encourage private investment in such efforts. For similar reasons, property rights could be assigned to those who undertake preservation of naturally occurring life forms for profit. To mention this possibility does not, however, constitute an endorsement of this approach. The ramifications of private property rights arising from "owning" rare species are multifaceted and would require considerable exploration.

For species in the wild, and for natural habitats and ecosystems, the idea of transferable private property rights offers little promise. Individual ownership of such things is impracticable for purely physical reasons and is so thoroughly inconsistent with tradition that it would enjoy little support. However, recent developments in the theory of incentive-compatible mechanisms suggest an alternative approach consistent with an individualist philosophy: the optimal "amount" of preservation and optimal individual taxes could be simultaneously determined in such a way that everyone would voluntarily consent to this optimal solution. I look with some favor on incentive-compatibility concepts. However, it would be simply untrue to claim that functional incentive-compatible devices capable of handling collective choices among large populations have actually been developed and tested. The day of incentive-compatible public choice procedures is yet to come if it comes at all.

The General Will, the Public Interest,
and the Social Function

Rousseau's version of social contract theory starts with the premise that all people are bound to the realization of equality, without which politics and justice are contradictions in terms. Governments expressing the "general will" must legislate only laws which are addressed to the common good of the society's members and which extend the same rights to and impose the same duties on each citizen.

Rousseau's philosophy foreshadows the modern emphasis on political equality and constitutional government. While Locke emphasizes sovereignty of the individual, Rousseau stresses the active participation of citizens in the political process as an indispensable condition for government by consent.

Modern public interest theories of government express faith that deliberative bodies are capable of identifying and interpreting the general will and of establishing policies and programs to implement it. To promote the general will over the interests of a powerful but selfish few, some considerable regulation of individual activities for the "public health, welfare, safety, and morals" may be justified. There is no need to minimize the scope of government, so long as government serves the interests of the general public. Thus, programs to promote economic activity, to rectify "market failure," and to promote equality of economic opportunity may all be seen as enhancing the general welfare and thus within the purview of government. Some conflict between individual aspirations and the general will is inevitable, but public interest theorists hope that individuals will be adequately protected by constitutional procedures and majority institutions.[19]

In mainstream economics, public interest and general will notions are expressed in the concept of the social welfare function, which provides a unique and unambiguous ranking of social states. Thus, assuming individual aspirations sometimes come into conflict, a social welfare function must specify the social consensus as to which kinds of individuals count how much. In practice the social welfare function is empirically elusive while there is much controversy, at the theoretical level, about the possibility of its existence.[20]

The public interest/social welfare function approach permits a wide range of regulatory initiatives, public investments, etc., for the purpose of ecosystem protection and species preservation. A case can be made that these things are legitimate uses of government authority in the public interest. While it is seldom possible to prove or disprove such a case objectively, public interest doctrine does not require this kind of proof. Rather, its premise is that, given a satisfactory political environment, those things that are approved through the public decision process will, *ipso facto*, be in the public interest.

Utilitarianism, the Potential Pareto-Improvement, and the Benefit Cost Criterion

Bentham, who was influenced by the social contract theorists and the early classical economists, popularized the notion that the primary human motivations are the pursuit of pleasure and the avoidance of pain. It is due to his influence that the word "utility" acquired two meanings: its customary meaning of usefulness and, in technical economics, its association with individual satisfactions and thus preferences. Initially, utility was thought to be, at least in principle, measurable on some cardinal scale.

While the difficulties of the cardinal concept of utility were mostly theoretical in the case of individual choice, they became crucial for political philosophy when the analysis was expanded to collective choice. Bentham thought that the proper criterion for collective choices was "the greatest good for the greatest number." That sounded fair enough, but entailed obvious difficulties in the quite plausible case where very great good to a few might be directly opposed to rather trivial good for many. If cardinal utilities could be summed across individuals, however, that could provide one solution to the problem: choose the alternative which maximizes the algebraic sum of utilities.

As it turned out, that avenue proved to be a dead end. Much later, around 1940, the concept of the potential Pareto-improvement emerged. *If* the gainers from some proposed change *could* compensate those who would otherwise lose, the potential Pareto-improvement criterion would find that change acceptable, even if compensation did not actually occur. Thus, the value indicators (WTP and WTA) would remain consistent with those that would emerge from exchange, but the Pareto-safety protection of voluntary exchange was held to be unnecessary. Here, at last, was an empirically applicable criterion by which the utilitarian concept of "the greatest good for the greatest number" could be implemented.[21]

The benefit cost criterion (that the benefits of change should exceed the costs, to whomsoever they accrue) is widely argued to be identical with the potential Pareto-improvement.[22] Benefit cost analysis (BCA) is an empirical test for potential Pareto-improvements.[23] This has an important implication. Values are determined with the status quo as the reference point—that is, as though each affected party had a right to retain his current level of welfare. Thus, gains (benefits) are valued at the beneficiary's willingness to pay for the benefit (i.e., WTP). Losses (costs) are valued at the amount of compensation which would induce losers to accept the loss voluntarily (i.e., WTA). The burden of proof is placed on the proposed change.

Nevertheless, the BC decision rule requires only that the sum of the value gained by beneficiaries exceed the sum of value lost by those who would be made worse-off by the proposed change. Thus, the benefit cost criterion provides no security for the individual. If this were to become the universal criterion for collective decisions, individual rights would be completely subordinated to the overall good of the collective. Resource reallocation could then legitimately proceed with government taking from the inefficient and giving to the efficient (as opposed to the exchange process in which the efficient purchase resources from the less efficient). Thus, philosophical individualists would be implacably opposed to such a collective rule. They favor efficiency, of course, but prefer that it emerge spontaneously from individual transactions rather than be deliberately imposed by governments armed with benefit cost analyses.

Public interest theorists would be little happier with such a decision rule, since it would enshrine a narrow, unidimensional concept of the public interest. Surely the general will is broader than that. Economists of the social welfare function school find the benefit cost criterion unduly restrictive. That criterion can be interpreted as one very specific social welfare function: one in which money endowments are substituted for utility and all dollars are weighted equally regardless of the circumstances of their owners. Few who defend the social welfare function concept would consider this its only acceptable form.

Thus the benefit cost criterion is an unacceptable social choice rule to philosophical individualists, to adherents of the public interest and social welfare function schools, and (for that matter) to a good many utilitarians. Perhaps that is why we never find the benefit cost criterion in use as the basic public decision rule. When it is used at all, it is invariably superimposed on the body of existing laws, which in the United States bows in the direction of individualism (e.g., by codifying private property rights) and the public interest (e.g., by regulating private activities so as to protect the public health, welfare, safety, and morals). Rather than determinants, benefits and costs become mere considerations in the choice from a set of alternatives which satisfy the basic law of the land.

Recently, the Office of the President, via Executive Order 12,291, has required that regulatory initiatives (newly proposed programs and those due for reauthorization) be subject to BCA, with a view to eliminating those which fail the benefit cost test.[24] This attempt to elevate the benefit cost criterion in the public decision process seems certain to precipitate legal battles between the legislative and executive branches of government and philosophical debates between adherents of the potential Pareto-improvement and public interest schools.

IV. ECONOMIC INFORMATION CONCERNING PRESERVATION ISSUES

The legitimacy of collective decision processes in a world where individual preferences come into conflict will always be subject to question. The limits of the various approaches have been defined, but no approach has been shown to be wholly satisfactory. Further, the benefit cost criterion, which has recently been promoted as a device to improve public choice in regulatory matters, has been shown to be an unsatisfactory social decision rule from the perspectives of both philosophical individualists and adherents of the public interest doctrine. At this point, we retreat from the search for the perfect societal choice criterion, and ask the less cosmic question: what can economic information contribute to policy decisions about the preservation of species?

Benefits, Costs, and Net Present Value

For many policy decisions regarding preservation of species, it is helpful to ask whether the net present value of a preservation action is projected to be positive, that is, whether its benefits are expected to exceed its costs (both appropriately discounted). BCA is an appropriate tool for answering this question. Individual valuations of the benefits and costs associated with preservation (Section II) are aggregated according to the potential Pareto-improvement rule (Section III). This kind of information could conceivably be used in any of the following ways:

(i) to rank preservation alternatives and establish priorities, or as a filter to eliminate proposals whose projected costs are greater than benefits (Section III).

(ii) as an information system organized to answer the question: would this action, if taken, contribute to national economic development broadly defined?

(iii) as an information base for evaluating the proposal in terms of its impact on regional economies, economic sectors, and economic classes. While the bottom line of BCA is a Yes/No answer as to whether a proposal's net effect on national economic development is expected to be positive, its information base is much broader and potentially useful in answering other kinds of questions.

(iv) as propaganda.[25] Certain industrial interests habitually present preservation issues as pitting the "hard-nosed realities" of economic development against "soft-headed environmentalist sentiment." BCAs, however, often show that aggregate WTP (or WTA, as the case may be) for amenities that would be provided by the pro-environment strategy is considerable. Not infrequently, the pro-environment alternative passes the benefit cost test. Given this situation, the propagandist dissemination of benefit cost data serves a useful public purpose. It changes the terms of the debate from the misleading binary idea, "real benefits versus sentiment," to the more pertinent spectrum focusing on "what kind of benefits, how much, and for whom?" It cannot hurt to establish the notion that commercial interests do not hold a monopoly on economic arguments. BCA (or for that matter, economic analysis more generally) is not inherently hostile to the preservation alternative.

Nevertheless, preservation of species problems are among those for which the impediments to high-quality BCA are greatest. The procedure for BCA, once the various alternative strategies have been defined, involves identifying the physical consequences of each strategy, evaluating these consequences economically (i.e., costs and benefits), and calculating the net present value of each strategy. Immediately, it can be seen that the economic analyst lives rather high on the information food chain. He/she functions by ingesting information about base line conditions and the consequences of change developed by practitioners of many other disciplines, and metabolizing it according to economic principles. When information about consequences is highly speculative (as it often is in preservation cases) the economist can do little to fill the information gap.

The basic value concepts, WTP for benefits and WTA for costs, are well defined and consistent with the logic of constrained choice. Nevertheless, measurement difficulties abound in problem situations (such as preservation issues) where nonexclusiveness, nonrivalry, and long time horizons are characteristic features. Three basic categories of valuation methods for dealing with these problems are available.[26] (1) Where well-functioning markets exist and the anticipated changes in quantity are small relative to the total quantities in the market, price is an adequate value indicator. Where demand and supply functions can be estimated from market data, these permit approximation of aggregate WTP and WTA even for large-scale quantitative changes. (2) For unpriced goods, there may exist closely related goods (complements or substitutes) whose markets can be analyzed to reveal information about the value of the unpriced good. (3) Hypothetical or experimental markets can be devised for unpriced goods and used in survey or experimental contexts to generate information on value.

Each of these methods has its limitations: (1) is inapplicable to many environmental goods; (2) is of broader but nevertheless limited applicability and requires various analytical assumptions which may introduce distortions; (3) is widely applicable, but the most readily implemented versions are not strictly immune to strategic manipulations on the part of respondents or experimental subjects. Given the present state of the art, only the methods of category (3) can be used to estimate option, quasi-option, and existence values (see Section II).

Those experienced in development and use of these various methods would claim that each, properly used, generates useful value information. On the other hand, each has its limitations and imperfections. Economists are not free, however, to manipulate value estimates in order to make their (purportedly scientific) conclusions consistent with their personal preferences. The relevant theory and methodological principles are sufficiently well developed to provide the basis for effective criticism. Researchers in the area are constantly pressured to attempt replication and other validation procedures.

In general, as one progresses from current use values to future (potential) use values, option values, and existence values, assessments of consequences and their economic significance become less reliable. The difficulties encountered in valuing a single environmental good are compounded when one moves to valuing the contributions of whole systems. Systems production concepts such as synergism play havoc with the notion of simply adding together the contributions of individual components. On the demand side also, recent work suggests that valuation of changes in complex systems cannot, in general, proceed via simple summation of the values (or demands) for individual components.[27]

The incidence of costs and benefits is, of course, imperfectly synchronized. BCA addresses this problem by reducing all benefit and cost entries to present value by discounting. This procedure recognizes the scarcity of capital and treats the interest rate (i.e., the price for using capital) as any other price. The logic of this procedure is impeccable in the context of evaluating an investment opportunity generically similar to the general run of investment opportunities. It is much less defensible for evaluating programs heavily weighted toward providing collective amenities and likely to influence many future generations.

On balance, I believe that BCA is likely to provide information useful to the public policy process. The formal procedure of BCA can be helpful for its own sake, since its rigorous structure disciplines investigators who might not otherwise remember to consider all of the relevant variables and their consequences. While BCA will inevitably rely on weakly substantiated data (about consequences and economic values) for some categories of costs and benefits, its capacity to mislead is minimized in an open, critical environment in which benefit cost documents are exposed to professional and public scrutiny.

These observations lead to two caveats. First, there is good reason to be suspicious of any agency which treats BCAs as internal documents, shielded from review by public and outside professionals. Such an agency could make honest mistakes. It could also seek to generate public support for programs favored by high-level administrators and those to whom they are responsive, on the (unexaminable and therefore potentially false) grounds that the favored programs provide a superior balance of benefits over costs. Second, inherent deficiencies in the data provide a pragmatic reason—in addition to the philosophical reasons mentioned in Section III—for objections to the use of the benefit cost criterion as a strict decision rule (a ranking device or a filter) in preservation issues.

Are economists interested in preservation issues best advised to persevere with BCA? Or should they seek some other conceptual framework as a basis for generating factual and prescriptive information? One approach to preservation issues, dubbed the Resources for the Future (RFF) approach,[28] explicitly reserves a central role for BCA. Every attempt should be made to measure the full range of preservation benefits, including the relatively speculative economic magnitudes associated with option, quasi-option, and existence values. Nevertheless, it seems almost certain, to adherents of the RFF approach, that major categories of preservation benefits will remain elusive. Any BC calculus will most likely underrepresent benefits relative to costs. Therefore, the RFF approach combines a requirement for formal and detailed BCA with a presumption that the preservation option should be given the benefit of any doubts remaining on completion of the BCA.

This position is not without merit. Still, there are preservationists who are offended by any obligation to document in economic terms the benefits of preservation; and there are some on the other side who are concerned that the "benefit of the doubt" clause licenses preservationists to put their thumbs on the preservation side of the scales.

Consequences and Opportunity Costs

An alternative approach suggested in the writings of the late S. V. Ciriacy-Wantrup and currently promoted by Richard Bishop is that of the *Safe Minimum Standard* (SMS). The SMS is defined as the minimum level of preservation which ensures survival. While Bishop is eloquent in elaborating the beneficial consequences of preservation (or the adverse consequences of failing to ensure the survival of species, habitats, and ecosystems), he resists any obligation to estimate benefits empirically. For those species not yet screened for commercial use, there is gross uncertainty about the consequences of loss (not mere risk, as the BCA approach implies).[29] Nevertheless, evidence is abundant that nature in general has provided inestimable subsistence, commercial, and amenity services to humankind. Continuing processes of exploration, investigation, and research sometimes reveal new potential uses for species previously thought "useless," and newly understood dependencies of "useful" species and valued ecosystems upon species not previously appreciated by humans. Bishop would take the incontrovertible evidence that species (in the plural) are of high and rising value as a sufficient basis for a presumption of positive value for any single species.

The BCA approach starts each case with a clean slate and painstakingly builds, from the ground up, a body of evidence about the benefits and costs of preservation. The SMS approach starts with a presumption that the maintenance of the SMS for any species is a positive good.[30] The empirical economic question is "Can we afford it?" or, more technically, "How high are the opportunity costs of satisfying the SMS?" The decision rule is to maintain the SMS unless the opportunity costs of so doing are intolerably high. The burden of proof is assigned to the case against maintaining the SMS.

The SMS approach avoids some of the pitfalls of formal BCA: e.g., the treatment of gross uncertainty as mere risk, the false appearance of precision in benefit estimation, and the problem of discounting. Its weakness is that, rather than providing answers, it redefines the question. Nevertheless, an appealing argument can be made that "Can we afford it?" with a presumption in favor of the SMS, unless the answer is a resounding NO, is the proper question.

Bishop has compiled several case studies of the opportunity costs of specific preservation decisions.[31] Opportunity costs of preserving the habitat of the California condor may exceed $3 million per year. The true costs may be less than that, however, because much of the income forgone is from oil and mineral extraction and these reserves may be worth almost as much in the ground as in production. The opportunity costs of preserving the snail darter were probably about zero at the time of the Supreme Court decision, and negative prior to that time (because the Tellico Dam was an economically unattractive project anyway).[32] Those for saving leopard lizard habitat are also quite low, since the known habitat is confined to one canyon in Ventura County, California, and the opportunity costs of preservation are limited to the value of forgone off-road vehicle recreation.

The vast majority of endangered species live in tropical environments, and what would be a tolerable opportunity cost in the context of the U.S. economy could well be an immense burden in many impoverished tropical countries. While this may often be a serious concern, Bishop reports a counterexample: the opportunity costs of preserving mountain gorilla habitat in low-income, tropical Rwanda seem to be tolerably low.

These various examples show that the opportunity costs of preservation are not always large, and when they are small that fact alone serves to simplify the decision. Not all preservation decisions are difficult.

The SMS approach differs from the BCA approach in that it starts with a presumption of positive benefits and denies an obligation to document benefits case-by-case. Apart from that, it has much in common with the BCA approach. Were the SMS to be treated as a strict decision rule, it would be a collective decision rule of the potential Pareto-improvement type. Decisions would be made on the grounds that (presumed) benefits exceed costs in the aggregate, with no particular respect for individual differences in preferences or exposure to costs. As an information system, it offers an argument addressed to societal benefits and a documentation of opportunity costs at the aggregate level. This is useful information to those concerned with broad societal objectives.

As propaganda, it is a bold ploy which may or may not pay off. Since many will draw attention to costs, the main propaganda function of BCA is to document the benefits of preservation, while insisting on a national accounting stance. The SMS approach maintains the national accounting approach to opportunity costs, but argues that benefits can be presumed rather than documented. If this argument is convincing, it has the considerable propagandistic advantage of shifting the burden of proof to opponents of preservation. The risk, however, is that it may fail to convince.

V. PRESERVATION PRIORITIES

Resource allocation is about establishing priorities in a context of scarcity. So, one would hope at the outset that taking preservation of species as a resource allocation problem would yield some guidance about preservation priorities. The intervening sections of this paper, however, introduce diverse considerations which tend, on balance, to dampen the hope for definitive conclusions about priorities. The "basic facts of the situation" are too elusive and the societal choice problems too intractable to permit easy answers.

Nevertheless, some useful things—preliminary to establishing priorities—can be said. Further, we can say some things about priorities based on observations about human preferences and on value-premises about legitimacy in collective choice.

Some Issues

Life Forms or Life Processes? As indicated (Section II), preservation can be interpreted as meaning the preservation of life forms or of life processes, i.e., the capacity for future evolution, speciation, and diversification. The preservation of some minimal life processes for at least several more human generations is surely a goal which would be accorded universal agreement. One would expect very broad agreement with the goal of preserving some minimal life processes "forever." One would expect little support for the idea of absolutely maximizing the capacity for future evolution, speciation, and diversification, since this would probably be enormously expensive.

If, as I conjecture, preferences are also focused on specific life forms, the options are viewed a little differently. Humans may decide to intervene to preserve favored species which appear otherwise doomed to extinction even if extinction could be considered part of the natural course of events. Maximizing the number of life forms preserved may be somewhat less costly than maximizing the scope for life processes, since opportunities exist for preserving some life forms independently of their "natural" habitats in zoos, botanical gardens, laboratories, and the like.

One would expect the options to be narrowed a little by these considerations. The minimal position would involve measures to ensure some continuing scope for life processes and protection of some favored life forms. The most aggressive feasible preservation strategy would fall short of maximizing the scope for life processes and would necessarily accept the demise of some species. Any viable strategy would need to recognize that positive preferences appear to exist for both preservation of species and preservation of the potential for new life forms to evolve.

Species or Habitats? A related question concerns whether preservation efforts should focus on species or habitats. Sometimes, there seems to be no choice: the species simply does not respond to known methods of propagation in zoos, botanical gardens, or laboratories. Often, however, there is a choice, and many opportunities exist to preserve species independently of their natural habitats. Such strategies are not appealing to those whose preferences are focused on ecosystems and the whole environment more than on particular species. Further, many of the amenity services provided by life forms are much enhanced when they can be observed in a more or less "natural" context. Nevertheless, several categories of value can be served by preservation in a man-made environment. Genetic information would be preserved, generating expected future use values, option values, quasi-option values, and those existence values associated with the life form itself (independently of "its place in the natural scheme of things"). The "zoos, gardens, and laboratories strategy" is surely a less preferred but also less expensive option, compared with a natural habitats strategy. It will likely be used in some cases, in a world of existential scarcity, at least to supplement an incomplete "habitats strategy."

The species or habitats question, however, is not directed entirely to the "artificial versus natural environments" issue. Assuming a prior choice to pursue preservation in natural environments, a question still arises of whether to focus attention and effort on the species or the habitat. A majority position seemed to emerge within the working group that a focus on habitat protection had clear advantages, since it would secure many related species as well as the target species, and preserve life processes as well as the target life form.[33] On the other hand, Richard Bishop argues that the species focus has its advantages, too.[34] Species serve as indicators of habitat problems. Protecting a species in the wild will concomitantly protect its habitat. Some prominent species have broader appeal than the more amorphous habitat and thus the species approach is a preferred form of pro-preservation propaganda. The substantive arguments tend to be mutually canceling and hence inconclusive. Therefore, the issue is perhaps one of effectiveness in propaganda: which approach has the most public appeal?

The Possibility of Priorities. Two issues suggest that the very possibility of setting priorities is itself in question: (1) are the production possibilities such that priorities are meaningful, and (2) given that preservation is necessarily a collective choice, what is the basis for this choice? Having discussed the second issue at length (sections II, III, and IV), I limit myself here to a reminder that problems remain. The first issue, however, requires discussion at this point.

The ultimate instrumentalist argument—that every life form must be preserved in order to ensure survival of the things people do care about and, ultimately, of humanity itself—tends to deny, or at least sharply restrict, the possibility of setting priorities. If the loss of one species would bring about *instant* ecocatastrophe, there is no basis for priorities among species or habitats; the loss of any ensures the immediate loss of all. If the loss of any species initiates processes leading to *eventual* ecocatastrophe, some choices can be made, by scheduling the disaster so as to maximize the net present value of aggregate human benefits between now and the end. The chosen strategy would allow the least-valued species and associations to go first, while prolonging the existence of the most-favored. Criteria for choice would be preference-based and would (given the inevitability of disaster in the end) tend to emphasize current raw material and amenity values.

The hope that humankind can, by choosing carefully, indefinitely prolong a high-quality existence on earth depends on the denial of the ultimate instrumentalist argument in its broadest form. It is true that species are interrelated and some species are threatened because their support network of associated species has been drastically altered. Nevertheless, it seems too easy and too unhelpful to insist that each is essential to the survival of all others. The meaningful setting of priorities beyond the mere scheduling of inevitable catastrophe requires more information about the patterns of dependency.

Priorities: Some Suggestions

Assume that it is possible to establish the linkages and patterns of dependency among species, and that some of the linkages are sufficiently weak to remove the possibility that the loss of any one makes inevitable the loss of all. On this basis, we can proceed to discuss priorities. We need some basis for making collective choices. I would suggest that individual valuations be summed to determine aggregate valuations, and that we prefer the alternatives whose aggregate value most exceeds its opportunity cost. That much is common to both the individualistic and potential Pareto-improvement approaches to providing nonrival goods. Followers of both approaches would agree as to what should ideally be preserved.[35]

At this point, species would be sorted into two groups: (1) those for which we know something about aggregate value and (2) those for which we have no empirical basis for valuation.

(1) For the first group, preservation priorities would be set using benefit cost logic, but not necessarily on the basis of formal BCA. The idea that an aggregate value far in excess of opportunity cost earns a species a high place on the priority list would be retained, and attempts to document benefits and costs would be encouraged. However, formalism in the use of the benefit cost criterion would be avoided, to avoid the various and sundry evils of such an approach: the obligation to document benefits regardless of the difficulties; the appearance of more precision than can honestly be claimed; and the problems with discounting and the coerciveness of the benefit cost criterion as a societal choice rule. This approach to setting priorities is, I believe, a near middle-ground position between the RFF and SMS approaches. It reduces the formalism of the RFF approach, while using more benefit information than would a strict SMS approach. Surely supporters of the SMS would not object to using benefit information when it is available; they object, I believe, to the obligation to document benefits when such information is neither reliable nor inexpensive to obtain.

Species likely to fall into the first group include: those which provide raw materials for human use; those which provide amenity services for human enjoyment; those which have good prospects of eventual recognition as providers of raw materials; those which are of known or expected usefulness in breeding programs to improve commercial varieties of plants and animals; and those which fulfill key roles in the ecosystems upon which all of the above depend for survival. In principle, no preference would be given to particular categories of value (i.e., raw materials values would not automatically count for more than amenity values; use values would not count for more than option, quasi-option, and existence values, except to the extent that the latter are more speculative than the former).

Some of the species in this first group would be assigned high priority for preservation "in the wild" and others for preservation in managed environments. Those whose known usefulness is in production of raw materials in managed environments or in maintenance of the gene pool, and those which seem to be losing ground in their natural habitats, would be assigned to the managed environments, provided they are known to have excellent prospects of surviving there. Those valued for habitat and ecosystem services, those which serve key roles in valued ecosystems, and those which do not adapt to managed environments would be prime candidates for preservation in minimally disturbed habitats.

(2) Species which do not receive a high priority rating in the first group would be reassigned to the second group, which then would include all species not already accorded a high priority. Among these, SMS logic would play the dominant role in establishing priorities. All species would be treated as having a positive but unknown expected value; implicitly, all would be treated as equally valuable. Priorities would be set on the basis of opportunity costs: preserve that package which includes the most species given the cost constraint. This is essentially a triage approach.

Those which can be preserved relatively inexpensively in managed environments are prime candidates for that kind of preservation. For preservation in relatively undisturbed habitats, "second group" species would score well if their habitat could be cheaply maintained or, alternatively, if preserving an expensive habitat would save an extraordinarily large number of life forms. Some species would fall into both subcategories, qualifying for preservation "for sure" in managed environments while getting to take their chances in preserved habitats.

In both the first and the second priority groups, species and subspecies which have many close relatives may rank a little lower than those which do not, for two reasons. First, closely related life forms may serve as substitutes in providing the beneficial services attributed to the threatened species. Second, assuming continued advances in genetics, the possibilities are more promising for developing adequate substitutes for such life forms through genetic engineering and artificial selection.

This scheme for establishing priorities does not address the question of the cut-off point: at what level of total costs can no further commitments to preservation be made? For reasons elaborated in earlier sections, it seems to me that there is no method for establishing the "right" level of collective expenditure on preservation which is both empirically tractable and satisfactory from the perspective of legitimacy in collective choice.

VI. Conclusions

Preservation of species has been treated as a resource allocation problem. Such an approach required, first, consideration of the logic of individual choice. Then, normative theories of societal choice were considered, since there is limited scope for effective preservation decisions at the individual level. I reached inconclusive results in the search for normative societal choice rules, after examining benefit cost analysis and the safe minimum standard approach as information frameworks in the context of preservation decisions. Finally, I offered some suggestions about preservation priorities. Some key points emerge.

On the Normative Economics of Societal Choice

(1) Individual preferences carry considerable normative weight. Individual choices, being influenced by preferences but also by information, endowments, and the existence and completeness of markets, have somewhat less normative import. The transition from individual to societal choice presents major, and perhaps insoluble, normative problems.

(2) Economics, which focuses upon, but also beyond, commercial and market behaviors, is not inherently hostile to preservation objectives.

(3) Economics, as an empirical discipline, lives near the top of the information food chain. While it is capable of generating elegant (if perhaps not universally acceptable) frameworks for setting priorities, empirical implementation requires enormous amounts of information about the physical and biological consequences of decisions, much of it currently unavailable.

On Benefit Cost Analysis and Its Alternatives

(4) The benefit cost criterion is directly derived from a theory of government. It implements the potential Pareto-improvement criterion which

is itself one specific utilitarian answer to the question of legitimacy in government.

(5) The benefit cost criterion, viewed as a strict decision rule, would be unacceptable to philosophical individualists and adherents of public interest theories of government. Individualists object to its lack of concern about nonconsensual and uncompensated losses to some individuals while it permits even larger aggregate gains for others. Public interest theorists object to its value system, which (i) aggregates individual valuations and thus implicitly accepts the existing distribution of endowments and (ii) implicitly denies any formulation of the public interest beyond the algebraic summation of private interests.

(6) Benefit cost analysis (BCA) may nevertheless serve a useful purpose in providing information for the public decision process.

(7) BCA includes all sources of benefits and costs which are consistent with viewing the preservation of species as a resource allocation problem. Included among benefits are raw materials and amenity uses (present and future), option and quasi-option values, and existence values that may arise from various altruistic motives.

(8) BCA is not inherently hostile to preservation objectives.

(9) Applications of BCA are impeded by inadequacies in information, both about the workings of the physical and biological world and about individual and aggregate valuations of possible outcomes of decisions. These impediments are especially severe in the application of BCA to preservation issues.

(10) A survey of the literature reveals two kinds of economic information systems about preservation: the RFF approach based on BCA with the benefit of the doubt (in the absence of complete information) accorded to the preservation alternative; and the SMS approach, which starts with a presumption that preservation is beneficial in every case and focuses on the opportunity costs of preservation.

(11) If executive agencies were formally to use BCA or the SMS approach in decision-making, the decision process would be well served by procedures that routinely open these information systems to public and professional scrutiny.

On Preservation Priorities

(12) There are some inherent conflicts between preservation of life forms and life processes. A coherent set of preservation objectives would address both aspects of preservation.

(13) Preservation of species in managed environments (zoos, botanical gardens, laboratories, etc.) may provide some, but seldom all, of the services provided by preservation in the "natural" habitat. Nevertheless,

the uses of artificial environments should not be overlooked. They may permit preservation of some species which would otherwise be lost, in at least two situations: (i) where preservation in the "natural" habitat is prohibitively costly, and (ii) where the species appears destined for extinction even though its habitat is not threatened.

(14) The setting of meaningful priorities is undermined by the argument that, since everything in the natural environment depends on everything else, failure to preserve each and every species will inevitably result in ecocatastrophe. If biological science can refute that argument and replace it with a more detailed but less sweeping understanding of interdependency, then it should be possible to set realistic preservation priorities.

(15) Drawing on the RFF and SMS approaches, a scheme for preservation priorities can be outlined. But this scheme, while it can suggest how a total "preservation budget" could be allocated, cannot tell us how much of society's resources should be devoted to preservation in the aggregate.

NOTES

1. The chapters in this volume by Regan, Sober, and Callicott deal with some of these viewpoints.
2. The case that choosing reflects learning processes, both conscious and unconscious, is stated vigorously by George J. Stigler and Gary S. Becker, in "De Gustibus Non Est Disputandum," *American Economic Review* 67 (1977): 76-90.
3. Compare Donald Regan's notion of "changeable preferences," this volume.
4. Ibid. For a viewpoint opposing Regan's, consider that of John R. Commons: "[The individual's] scheme of apportioning resources is his plan of life. It is his scheme of both economy and ethics." *Legal Foundations of Capitalism* (Madison: University of Wisconsin Press, 1959), p. 57.
5. Alan Randall, "The Problem of Market Failure," *Natural Resources Journal* 23 (1983): 131-48.
6. There is a considerable literature expanding on this perspective. One good review is G. Rausser, E. Lichtenberg, and R. Lattimore, "Developments in Theory and Empirical Application of Endogenous Government Behavior," in *New Directions in Economic Modeling and Forecasting in U.S. Agriculture*, edited by G. Rausser (Amsterdam: North Holland, 1982).
7. See D. Brookshire, A. Randall, and J. Stoll, "Valuing Increments and Decrements in Natural Resource Service Flows," *American Journal of Agricultural Economics* 62 (1980): 748-88; and A. Randall and J. Stoll, "Consumer's Surplus in Commodity Space," *American Economic Review* 70 (1980): 449-55, for definitions of the value measures, theoretical justification for their

use, and a discussion of their relationship with other value measures such as price and consumer's surplus.

8. See Richard C. Bishop, "Option Value: An Exposition and Extension," *Land Economics* 58 (1982): 1-5; and Daniel A. Graham, "Cost Benefit Analysis Under Uncertainty," *American Economic Review* 71 (1981): 715-25.

9. K. J. Arrow and A. C. Fisher, "Environmental Preservation, Uncertainty, and Irreversibility," *Quarterly Journal of Economics* 55 (1974): 313-19.

10. These points are more completely developed in A. Randall and J. Stoll, "Existence Value in a Total Valuation Framework," in *Managing Air Quality and Visual Resources at National Parks and Wilderness Areas,* edited by R. D. Rowe and L. G. Chesnut (Boulder, Colo.: Westview Press, 1983), pp. 265-74.

11. Compare the position of J. V. Krutilla, "Conservation Reconsidered," *American Economic Review* 57 (1967): 777-86.

12. Regan, this volume.

13. Slobodkin, Vermeij, and Norton, in this volume, discuss some of what is known.

14. See the chapter by Geerat Vermeij, this volume.

15. For a more detailed discussion of this point, see Norton's discussion of the "contributory value" of species, this volume, Chapter 5.

16. James M. Buchanan, *Freedom in Constitutional Contract* (College Station: Texas A and M University Press, 1977).

17. It is recognized that Pareto-safety protects the status quo and is just only to the extent that the status quo itself is just. Thus, Buchanan in *Freedom in Constitutional Contract* has explored a two-stage constitutional-contractarian process which employs a Rawlsian mechanism (see John Rawls, *A Theory of Justice,* Cambridge: Harvard University Press, 1971) at the first stage and Pareto-safety in the second stage.

18. See, e.g., T. N. Tideman and G. Tullock, "A New and Superior Process for Making Public Choices," *Journal of Political Economy* 84 (1976): 1145-59; and T. E. Groves and J. Ledyard, "Optimal Allocation of Public Goods: A Solution to the Free-Rider Problem," *Econometrica* 45 (1977): 783-809.

19. Here, the conflict between public interest theorists and individualists comes to a head. Individualists are reluctant to attribute any moral force to the outcomes of majority decisions, since they tend to view majorities as merely successful coalitions of individuals imposing their will on losing coalitions.

20. K. R. Arrow has denied the possibility of a unique and logically satisfactory social decision rule which satisfied a minimal set of democratic precepts. See his *Social Choice and Individual Values* (New York: Wiley, 1951). There has been much debate as to whether the social welfare function is or is not conceptually equivalent to the kind of decision rule Arrow analyzed. But, suffice it to say that the burden of proof seems to have shifted to the supporters of the social welfare function concept.

21. While the potential Pareto-improvement is a utilitarian criterion, it is merely

one of several solutions in the utilitarian tradition to the problem of collective choice.

22. See Randall and Stoll, "Consumer's Surplus"; Brookshire, Randall, and Stoll, "Valuing Increments"; and E. J. Mishan, *Cost Benefit Analysis* (London: George Allen and Unwin, 1971 and 1976).

23. Occasionally, one encounters the claim that benefit cost analysis, being merely an interloper from the arena of private and corporate decision-making, has no proper place in (or is strictly irrelevant to) public decision-making. See Mark Sagoff, "Economic Theory and Economic Law," *Michigan Law Review* 79 (1981): 1393-1419. As the preceding discussion has made clear, the benefit cost criterion emerges from one version of the utilitarian approach to the "legitimacy of authority" question. There can be no question that the benefit cost criterion is a solution to the problem of collective choice; the question is whether it is a good solution.

24. See *Federal Register* 46:33 (February 19, 1981): 13193.

25. The dictionary definition of propaganda is "material intended to persuade." The term propaganda does not directly address the quality or veracity of such information. My point is that accurate information may serve as propaganda. It may well be the best kind, since it is much harder for opponents to refute or deny.

26. There are many useful discussions of these methods. See, e.g., A. Myrick Freeman III, *The Benefits of Environmental Improvement: Theory and Practice* (Baltimore: Johns Hopkins University Press, 1979), the "Symposium on Environmental Management: The Policy Perspective," *Natural Resources Journal* 23 (1983), and G. L. Peterson and A. Randall, eds., *The Valuation of Wildlife Benefits* (Boulder, Colo.: Westview Press, 1984).

27. John P. Hoehn and A. Randall, "Aggregation and Disaggregation of Program Benefits in a Complex Policy Environment," presented to annual conference, American Agricultural Economics Association, Logan, Utah, August 1-3, 1982, and abstracted in *American Journal of Agricultural Economics* 64 (1982): 1079.

28. Richard Bishop coined that label to characterize the approach of John Krutilla, V. Kerry Smith, and Anthony Fisher (all of whom are or have been associated with Resources for the Future, Inc., Washington, D.C.) and others who have similar perspectives. See his "Endangered Species and Uncertainty: The Economics of a Safe Minimum Standard," *American Journal of Agricultural Economics* 60 (1978): 10-18.

29. The concept of risk assumes that enough is known to permit specification of a pay-off matrix with probabilities assigned to each cell. Uncertainty is relevant when the available information is inadequate to support that kind of analysis.

30. Bishop, "Endangered Species and Uncertainty," obtains this result with a game theoretic model. The SMS is the "modified minimax" solution.

31. Richard Bishop, "Endangered Species: An Economic Perspective," *Trans-*

actions of the 45th North American Wildlife and Natural Resources Conference (1980): 208-18.

32. See, also, A. Randall, *Resource Economics* (Columbus, Ohio: Grid Publishing, 1981), chap. 18.
33. See Leitzell's chapter and Norton's Epilogue, this volume.
34. Personal communication, August 20, 1982.
35. The two approaches diverge, of course, when the focus turns to individual protections. The individualistic approach would seek unanimous approval of the preferred program via optimal taxes determined with incentive-compatible mechanisms. The potential Pareto-improvement approach would tolerate uncompensated losses for some, so long as they were algebraically offset by gains to others.

5

On the Inherent Danger of
Undervaluing Species

BRYAN G. NORTON

I. INTRODUCTION

Many preservationists have assumed that the question they should address in attempting to affect policy is: How should one place an economic
or utilitarian value on a particular species?[1] My purpose here is to ask
and begin to answer a related but different question: What is the utilitarian value of a species, independent of its individual or populational
characteristics? The reformulation is significant. The first question emphasizes actual or potential commercial and other values which can be
assigned to some particular target species. It is answered by reference to
the populational characteristics of the species and/or to the physiological,
chemical, or ecological properties of its members.

My question is more general. Are there reasons for saying that, in
comparing a world with *n* species and another with the same *n* species
plus one more, the latter world is preferable from a purely human perspective? Answers to the more usual question, of course, provide instances
of human uses of other species which may, in a sense, underlie and
constitute the value of species in general. But the proposed question
encourages its asker to extrapolate beyond known and demonstrable
values of particular species.

The question does not imply that all species have equal utilitarian value.
The particular characteristics of specific species may well increase their
value or the probability that they will prove useful. The point of the
question is to ascertain the degree, if any, to which each and every species
has a value, independent of those special characteristics which give it
particular uses and potentials.

Many species preservationists are uneasy with any purely utilitarian approach to assigning values to species. At least a part of that uneasiness originates in the narrowness of the standard approach to assigning values to particular species. That approach presupposes that each species must "justify itself" in purely human and economic terms. The burden of proof lies on the preservationist to find or create values for a species or else regard it as valueless. This uneasiness motivated David Ehrenfeld's argument that natural objects should be treated as "non-resources."

> Nature . . . is seen as a "gigantic toolshed," and this is an accurate metaphor because it implies that everything that is not a tool or a raw material is probably refuse. This attitude, nearly universal in our time, creates a terrible dilemma for the conservationists or for anyone who believes of Nature as Goethe did, that "each of her creations had its own being, each represents a special concept, yet together they are one." The difficulty is that the humanistic world accepts the conservation of Nature only piecemeal and at a price. There must be a *logical, practical* reason for saving each and every part of the natural world that we wish to preserve. And the dilemma arises on the increasingly frequent occasions when we encounter a threatened part of Nature but can find no rational reason for keeping it.[2]

While I do not here intend to deny the existence of values not tied to human utilities,[3] I will argue that one need not advocate intrinsic value for nature in order to avoid the "conservationist's dilemma" whereby a species must be seen either as commercially useful or as useless.[4] By demonstrating that natural diversity, in and of itself, has utilitarian value, I will show that each and every species has prima facie utilitarian value which does not depend upon discovering some economic or industrial use for it. Thus one can avoid the undervaluing of species attendant upon "piecemeal" justifications, without transcending the realm of human utilities. The effect will be to shift the burden of proof in controversies between developmental and preservationist perspectives. If each species is valuable, a presumption exists against endangering any. Further, the value of each species is considerable, since we face the danger of a reduction in biological diversity, unprecedented in rapidity if not in scope, over the next few decades.

II. The Concept of Diversity and Species Preservation

Species diversity is necessary, but not sufficient, for complexity. Complexity implies numerous interactions of various kinds. A diverse natural ecosystem will also be relatively complex, as diverse elements can co-exist only if complex mechanisms have evolved to regulate those inter-actions. (Zoos are diverse without being complex; organization is im-posed on artificial ecosystems by the application of energy and infor-mation from outside them.) Since diversity in nature presupposes a degree of complexity and greater diversity implies greater complexity, I follow most researchers in using "diversity" to imply significant degrees of com-plexity as well.

Considerable controversy has arisen on this point. Some researchers believe that mere species counts, if not corrected for evenness (relative abundances), are not a useful indicator of complexity and that complexity is the ecologically significant characteristic. This indeterminacy has led Stuart Hurlbert to declare species diversity a "nonconcept" and to ad-vocate its replacement by a cluster of mathematical measures.[5] Fortu-nately, this controversy can be avoided in the present context, as species counts (species "richness" in Hurlbert's terminology) will prove to be the relevant characteristic for our purposes.

R. H. Whittaker and Robert MacArthur have distinguished three gen-eral types of diversity.[6] *Within-habitat* diversity refers to the number of species and to the number and complexity of interactions within an ecosystem. It presupposes (more or less artificially) that nature can be compartmentalized into systems which can be isolated from external interactions. *Between-habitat* diversity refers to the degree of dissimilarity between these closed systems. Thus, while species counts may be a rea-sonable measure of within-habitat diversity (especially if they are cor-rected for evenness), they are inapplicable to between-habitat diversity.[7] The latter concept must measure and compare dissimilarities in the types of species, the mix of species, and the types of interactions between them, as these exist in different systems. *Total diversity* refers to the variety of species existing in a geographical area. Because this concept need not presuppose sharp boundaries between ecosystems and, consequently, need not measure the types and complexity of interactions between spe-cies, species counts provide an adequate measure of it.[8] Further, geo-graphical boundaries can be set arbitrarily, on the basis of political boundaries or of gross geological features, and do not require knowledge of energy flows and species interrelationships as does the determination of ecosystem boundaries. Total diversity is, then, a function of the other two types of diversity in an area, but it is definitionally independent of them.

Besides being easily measured and avoiding presuppositions about eco-system boundaries, total diversity focuses on extirpation of species from an entire area rather than from particular systems. The extirpation of a species from a particular ecosystem need not be cause for alarm—it may be due only to a natural successional change in that system and involve a perfectly natural shift in species composition. Such extirpations usually indicate no problem as long as there are other populations of that species in the geographic locale.[9] The species remains available for colonization back into the area if some disturbance returns the system to an earlier stage of succession, making that species again a viable and appropriate part of that system.[10] But if a species disappears from more and more systems in an area, this trend indicates a loss of diversity and may justify concern regarding that species.

Local extinctions (from single habitats or systems) may not be grounds for concern. Yet it would be unfortunate if policies were set only to avoid global extinctions. Concern for global extinction alone can lead to the following reasoning: "It is expensive and/or inconvenient to protect the remaining populations of grizzly bears in the forty-eight states. Since larger populations exist in Alaska and Canada, we may as well concentrate our efforts on preserving those." This reasoning is very shortsighted. Besides providing the best possible insurance against an unavoidable calamity (such as an epidemic of a fatal disease in the large population), isolated populations keep the gene pool of the species diverse and flexible. The species is better prepared to react to changing environmental pressures and, in the very long run, is in much better position to undergo further speciation. Concern for total diversity, which measures species persistence in areas, continents, and the world as a whole thus provides the proper focus: the elimination of a species from its historical range is reason for concern.

For these reasons, I suggest that total diversity is the concept of proper concern for species preservationists.[11] Each species in an area can be viewed as a unit of total diversity. Not surprisingly, then, advocates of protecting species from extinction naturally fall into discussions of total diversity. Thus Aldo Leopold remarks that "to keep every cog and wheel is the first precaution of intelligent tinkering."[12] Likewise, in his chapter on "The Conservation of Variety," Charles S. Elton, often cited as the first to advocate conservation measures based upon the diversity/stability hypothesis, employs examples which strongly indicate an interest in total, not within-habitat, diversity. His references to the advantages of species variety as embodied in a patchwork environment composed of mixed fields, hedgerows, and roadside verges clearly indicate concern for total diversity.[13]

III. Diversity as Self-Augmenting

I now pose the central question of this paper: What human, utilitarian values are served by any species viewed as a unit of area-wide total diversity? Five arguments to be developed below establish, by increments, the considerable value of biological diversity and, in turn, of the specific species which constitute it. All five utilize one or the other of two premises derived from theoretical ecology: diversity augments diversity, and diminutions of diversity cause further such diminutions. That is, diversity is self-augmenting, and (barring serious disturbances) there is a positive diversity spiral—every increase in diversity provides opportunities and encouragement for further increases. Likewise, every loss in diversity causes further losses. Severe natural or human-caused habitat and ecosystem disturbances can generate a downward spiral in diversity. This, too, is a self-augmenting and accelerating process which is very difficult to reverse.

Let me first explain briefly how and why these principles are true. Diversity augments diversity both in ecological time through invasion and colonization and in evolutionary time through competitive pressures, co-adaptation, and natural selection. Since the former process contributes to the latter, it will be useful to begin with it.

Ecological Time

Our concern is with how total diversity protects and encourages total diversity. However, since an area's total diversity is a product of its within-habitat and between-habitat diversity, all three concepts are involved in an explanation of the relevant processes. Although traditional, strong theories of ecological succession have been for some time in ill favor among most ecologists,[14] it is still agreed that, following a disturbance, predictable patterns govern invasion and colonization by plants and animals of the disturbed area. Opportunist species, which are well suited to colonize spaces opened by disturbances, are usually not good competitors in the long run. So, it can be expected that species with fast growth rates, short life spans, and high dispersal rates will provide the dominant structure immediately after a disturbance. Later, long-lived plants which divert energy from propagation for the development of biomass will predominate. The picture which emerges from this minimal theory of succession is of a patchy landscape—competition will often be affected by random factors of dispersal, disturbances of varied magnitudes will continue, and the result will be a harlequin environment varying in species make-up across space and time.[15] The vegetational structure of the community will, in turn, determine which animal species are likely to exist and dominate in various patches.

Species will colonize disturbed areas and compete for niche space with species already present and with other invaders. Total diversity in such a mosaic of habitats and micro-habitats is likely to be quite high,[16] even though species may be subject to frequent local extirpations. Simon Levin describes these processes illuminatingly:

> . . . movement between patches is generally thought of as having a homogenizing effect, but this need not be the case. Fugitive strategies maintain species in the community through spatio-temporal patterns which may initiate, aggravate, and capitalize upon phase differences. The fugitive is doomed locally, but survives globally by a balance between dispersal ability and competitive or escape ability. It is only a slight extension of this notion to replace direct dispersal from patch to patch by indirect (and delayed) dispersal. . . . When this generalization is made, the fugitive becomes the rule rather than the exception, since most species are, to some degree, locally ephemeral.[17]

An area's total diversity is constituted by and at the same time maintains and increases this mosaic of community structure, because the availability of seeds, propagules, and migrating animals is an essential part of ecosystem development.

It is impossible to overemphasize the importance of a large and diverse pool of species available to colonize open spaces for the development of community structure. These effects are not limited to early stages of succession. Martin Cody, noticing great variation in between-habitat diversity in cross-continent comparisons of otherwise similar areas, speculated that an important factor was "the relative accessibility of each habitat to a species pool of more or less appropriate colonists from elsewhere on the continent."[18] Cody found that approximately 50 percent of the variation in between-habitat diversity was explained by differences in accessibility.[19]

Thus, the total diversity of an area provides the pool of competitors for niches in developing ecosystems. The larger this pool, the more likely it is that the system will evolve into a complex, highly interrelated system. And complex, highly interrelated systems provide more niche opportunities for new species. Over time, interspecific dependencies, both of predation and mutualism, will evolve. Further, interspecific competition often aids in avoiding competitive exclusion, as predators concentrate on the competitively advantaged species on any given trophic level. Thus, total diversity plays a key role in the development of ecosystem structure through ecological time. That structure, in turn, provides opportunities for more species to survive and thereby increases total diversity further. Therefore diversity augments diversity in a continuing upward spiral.

Evolutionary Time

These same processes explain, indirectly, how diversity is also self-augmenting in evolutionary time. R. H. Whittaker describes the general process as follows:

> Consider . . . the niche space for a group of organisms in a community. Along each axis of that space the number of species tends to increase in evolutionary time as additional species enter the community, fit themselves in between other species along the axis, and increase the packing of species along axes. Species can also be added as specialists on marginal resources, and they can be added by the evolution of new resource gradients and species adapted to utilizing them. Thus the evolution of modern flowering plants in the Tropics has made available new resources of nectar and fruits, and bird species have evolved that rely on these. Considered for a given group of organisms, diversity increases through evolutionary time by the "lengthening" of niche axes and the packing of more species with narrower niches along niche axes, and by the addition of new axes—by the "expansion" and complication of the niche space.
>
> Consider next two interacting groups of organisms, such as plants and grazing animals. The greater the diversity of plant species, the greater the variety of resources for the grazers. . . . But the influence works in the opposite direction also. The grazing animals may increase diversity of the plant community by preventing any plant species from becoming too strongly dominant. . . . Furthermore, a species that controls the population of another species is an important part of the latter's niche. Increased numbers of grazing species may imply increased numbers of niches distinguished by these as population controls, for the grazed species. Interacting groups of species can thus each facilitate the other's increase in diversity. We should expect the same effect between predators and animal prey, and between plants and symbiotic fungi.
>
> We can thus say that diversity begets diversity. Species diversity is a self-augmenting evolutionary phenomenon; evolution of diversity makes possible further evolution of diversity.[20]

It is important to note that, in Whittaker's description, the entire process gains its initial and ongoing impetus "as additional species enter the community." Area-wide, total diversity is, then, a prerequisite for this process. Greater degrees of total diversity will intensify and accelerate the process.

Diminutions in diversity affect the spiral in reverse. Losses in diversity beget further losses and the upward diversity spiral will be slowed and eventually reversed if natural and/or human-caused disturbances are severe and continued.

If a species goes extinct, other species which interact with it and depend upon it are, in turn, threatened. Peter Raven has asserted that, for every plant species which becomes extinct, fifteen animal species can be expected to follow.[21] In general, species which have evolved exclusive relationships, feeding or otherwise, with other species are more likely to suffer extinction, because their inherent probability of extinction is increased by the likelihood of the species on which they depend becoming extinct.[22] Even species which have no exclusive dependencies derive greater risk from the importance of any relationships they do have. Thus systems characterized by high degrees of complexity, interdependence among species, and large numbers of specialized species—those with the most biological diversity—are precisely those which suffer most from disturbances sufficient to extirpate one or more species from them. The simplification of systems relaxes the pressures of natural selection on remaining species, lessening the need to develop specializations, so the reversal of the diversity spiral operates in evolutionary as well as ecological time.

These related theories provide the basis for five arguments for the protection of biological diversity.

IV. The Contributory Value of Species

Argument 1

That diversity produces diversity does not, of course, imply that diversity is valuable. To show that, one must show that diversity is preferable to simplicity—the central issue in species preservation. Here the familiar species-by-species accounts of the value of species become relevant. Given the unquestioned presumption that some as yet unidentified and/or unexamined species will prove useful in the future, increments in diversity increase the likelihood of utilitarian benefits to man. If diversity contributes to diversity and if it can be assumed that some significant subset of any random collection of species will prove useful to humans, then all species can be assumed to be of some value, direct or indirect. Even species that have been unsuccessfully examined for human uses are still useful, because they contribute to increases in diversity which, in turn, contribute to the generation of more species. Some of those species will turn out to be useful to humans and the species that were of no direct use will prove useful indirectly.

Even species that fail in the competition for niche space in a particular ecosystem have made an important contribution to the structure of that system. By intensifying the competition for niche space, they have exercised adaptational pressure on their superior competitors, caused further specializations, and developed more tightly packed and more highly integrated species relationships. Consequently, even the losers in such competitions are important, and ecosystem development and total diversity are served, both directly and indirectly, by the continued existence of a species in an area.

Argument 2

A second, similar argument for protecting species emerges from the claim that diminutions in diversity beget further diminutions. If a "useless" species is lost from an area, other species that depend upon it will, likewise, be lost. Scientific understanding of ecosystems is too limited even to begin to list interdependencies among species, so it is impossible to predict which species will be included in the cascading effect of extinctions resulting from an initial extinction. Any decision to extinguish a species or to allow one to go extinct moves toward a policy that places higher priority on economic efficiency and development than on protecting biological diversity. Such a policy accelerates a spiraling course toward a less complex and diverse biological world. Since species exist in interrelated systems which have taken aeons to evolve by processes humans are unable to duplicate, the choice to extirpate a species from those systems inevitably weakens them.

To see this, it is useful to distinguish what might be called first-order dependencies from higher-order dependencies of species. Species a has a first-order dependency upon species b if b's extirpation immediately removes an element of a's environment essential to its survival. Even if (contrary to fact) one knew all the first-order dependency relationships within an ecosystem and could use this information to compute the other species losses caused by an extirpation, there would remain the further, higher-order dependencies about which ecologists know next to nothing. These are often very subtle, involving small changes in the habitats of other species that make the environment slightly less suitable and begin a slow deterioration in its population. Often, these relationships will be complex. For example, the population of one species may be reduced because extirpation of another lessens but does not eliminate its food supply. But a third species may have a first-order dependency upon the species with diminished population to such an extent that it cannot be supported at all by the smaller population. Or, the diminished population might be lost years later because it has become more susceptible to random environmental fluctuations. The full effects of extinctions, especially multiple ones, may not be felt for centuries.[23] Ecologists cannot delineate all of these relationships and cannot provide economists with the information necessary to compute the economic effects of a species loss. A computation of the true value of a species to the human race, over an unlimited time frame, must include all the event's unpredictable ramifications—a task to which neither ecologists nor economists are equal. But it is possible to infer that the value is extremely large. A decision to limit value computations to brief time frames, such as thirty years, is to sacrifice the future options of the human race to short-term economic gain.

Because species exist in ecosystems too complex for current human understanding, the loss of a species is accurately viewed as a first step in a process of ecosystem simplification. Viewed in this manner, it is more likely that the gravity of the losses will be recognized, even if it is impossible to place dollar figures on them.

The first two arguments establish that the prima facie value of a species is significant. When higher-order dependencies of other species on extinguished ones are taken into account, it becomes almost certain that some species eventually lost in the processes begun by the extinction of just one species will prove to be of substantial worth to humans.

V. The Magnitude of the Downward Spiral and Zero-Infinity Dilemmas

It is possible that skeptics might respond that, while some useful species are lost as a direct or indirect result of an extinction, there are so many species and we have so little research time to examine them that losing or saving a few more useful ones is not very significant.[24] The skeptic might say that the protection of a species is only a beneficent act.[25] It is generally believed that the obligation to be beneficent (as in giving to the poor) is weaker than the obligation not to harm someone. Where there is a wealth of species, the protection of one might be considered merely an act of beneficence—the obligation to provide more of a good thing to future individuals who will already be well blessed. This weaker form of obligation might be thought to be overridden easily by matters of convenience and short-term economic gain. That a great deal of diversity is essential, in the long run, for human health and well-being does not, in this view, entail that diversity must be maximized. The obligation to protect a considerable amount of diversity would be seen as much stronger than the obligation to protect each and every species.

A first response to this skeptical objection is that it seems insensitive to the cascading effect of species losses—one is virtually never discussing the loss of just one species. The loss of a species represents the beginning of a trend.

Note that the skeptical objection carries differential weight, depending upon how one views the present situation. If biological diversity remained in an upward spiral, then the loss of a species would merely slow the acceleration of the spiral and the obligation to preserve a species could be considered an obligation of mere beneficence, rather than the stronger obligation to avoid harming future individuals. The skeptical objection rests upon an empirical assumption that the biological world is, in fact, still increasing in diversity. But this empirical assertion is false. The current rate of extinction is already far in excess of the historical rate of speciation.[26] The loss of a species now is not merely a slowing of the spiral, as it would have been a few centuries ago.

Further, the accelerating nature of the downward spiral implies that every species loss is more important than the one before. Norman Myers estimates that the extinction rate is accelerating as follows: there was only one extinction per 1000 years even during the elimination of the dinosaurs.[27] Since man has exerted a force upon extinction rates beginning about 50,000 years ago, these rates have increased dramatically.[28] Anthropogenic species extinctions from 1600 A.D. to 1900 A.D. probably totaled about 75 species.[29] But this rate has increased alarmingly. For much of this century, the rate was one human-caused extinction per year, but since 1960 it has skyrocketed; estimates range from 1000 species lost per year to even higher figures.[30] If, as many experts have predicted, 1 million species will be extinguished in the final quarter of this century, this would represent an average of 40,000 lost per year.[31] This figure would be much lower currently, but would be expected to grow to a rate where one species is lost each hour.[32] While these figures are not based upon precise data, the rate is unquestionably accelerating and any rate in this range would be unprecedented in the known history of the planet.[33] Speciation rates vary according to species and according to time, but are certainly much lower than current and projected extinction rates.[34] A downward spiral in biological diversity has most certainly begun and it is accelerating at an awesome rate. It is these alarming figures which have led to projections that as many as one-fourth of the currently extant species will be lost by the year 2000.[35]

Biotic communities all over the world are becoming progressively impoverished, even when local extirpations do not involve worldwide extinctions. Reintroductions of rare species into their historical ranges almost never work and, when they do, are extremely costly.[36] As we approach the twenty-first century, many species which are not actually extinct will have such limited ranges that their susceptibility to extinction will be very great. And, even when strong populations remain in a vastly truncated range, the benefits of living in a more diverse area are lost to humans in the areas where the species has been extirpated. As frightening as are the current and projected rates of worldwide extinctions (diminutions of diversity worldwide computed over evolutionary time), the rates of local habitat destruction and ecosystem simplification (diminutions of diversity in localities computed across ecological time) are even more so. This skeptical objection and the response to it therefore set the stage for another argument for protecting species.

Argument 3

A third argument emphasizes that humans, because they exist at the end of various food chains and depend upon other species in many, not always well understood, ways, are especially vulnerable to threats to their own existence. Increases in the global rate of extinctions increase the vulnerability of the human species to extinction.

Paul and Anne Ehrlich preface their book *Extinction* with a parable called "The Rivet Poppers." A person entering an airplane for a flight notices a workman prying rivets out of the wings. The workman explains that the rivets can be sold for $2 each and that this subsidy allows efficient service at affordable prices. When questioned about the safety of the practice the workman merely replies that it must be safe as no wings have yet fallen off planes, even after successive rounds of rivet-popping.[37] The analogy should be obvious. No particular species extinction of the many caused by man has resulted in a major disaster. So, the evidence from past cases points to there being no disasters resulting. The problem of diminutions in diversity embodies characteristics associated with other environmental risks with low probabilities and high consequences, sometimes called "zero-infinity dilemmas." If too many species are lost by increments from an ecosystem, an area, or the worldwide biotic community, a catastrophic ecosystem breakdown could occur. The risk of any particular extinction having catastrophic effects might be quite small. Assigning of economic values to such risks is notoriously difficult.[38]

Such risks involve nine characteristics, as listed by Talbot Page: (1) Ignorance of mechanism: we lack knowledge about how the disaster would occur; (2) Potential for catastrophic costs: a disaster could be very costly in economic terms or even in human lives; (3) Relatively modest benefits: the advantages are small and incremental, rather than great and sudden; (4) Low subjective probability: the disastrous effect is unlikely; (5) Internal transfer of benefits: the benefits associated with the risk are transferred through markets and reflected in product prices; (6) External transfer of costs: the risk's adverse effects exist in the environment; they are not expressed through the market (they are economic externalities); (7) Collective risk: the risk is faced by many individuals; (8) Latency: there is an extended delay between the action creating the hazard and the manifestation of its effect; (9) Irreversibility: once the effect occurs it is, in a practical sense, irreversible.[39]

Actually, the rivet-popping parable embodies an inductive fallacy: the assumption that the failure of a low-probability event to occur decreases the likelihood that it will occur in the future. Two kinds of situations must be distinguished here. In some situations, such as catastrophic core meltdowns of nuclear reactors, the individual occurrences are independent. The absence of a meltdown to date can be considered (very weak) confirmation of a low probability assignment to future meltdowns. In others cases, occurrences are not independent, as when an urn contains one red ball and 1,000 white ones, and each failure to draw a red ball increases the probability that the next will be a red one. Assuming, as most ecologists would, that there is some minimal number of species such that, if extinctions diminished the stock of species below that minimum an ecological disaster would occur, species extinctions represent dependent occurrences. Each species extinction increases, however slightly, the probability that the next will prove disastrous. Hence, the rivet-popping parable exhibits the false assumption that dependent occurrences are independent.

In spite of this fact, there are important logical affinities between the rivet-popping analogy and zero-infinity dilemmas. Because there are so many species, the likelihood that the next extinction will produce a disaster is very low. But this low probability of disaster derives from the fact that ecosystems contain considerable redundancy. This redundancy has developed over aeons of time by the processes described in Section III. When human disturbances begin to reverse those processes, each extinction increases the very small probability that the next one will prove disastrous. That this probability is initially very low (a feature shared with other zero-infinity dilemmas) encourages the fallacy that nondisastrous extinctions are evidence that all will be so. Once we see through the fallacy, then, we see that the eradication of species involves a zero-infinity dilemma. There is a very small probability that the next species extinction might involve a serious disaster. (I assume that any occurrence involving a large number of human casualties and/or a breakdown of the ecological and economic system of a large populated area would count as a disaster.)

Ecologists warn that species extinctions, simplifications of systems, and consequent decreases in total diversity carry at least some risk of bringing about an environmental catastrophe. Orie Loucks has argued, using reasoning related to mine about the effects of diminutions in diversity, that over-management of ecosystems can lead to their failure to rejuvenate themselves. Diversity progressively deteriorates and the productive potential of entire areas eventually collapses. The progressive transfer of all systems in an area from a natural to a managed state interrupts the natural cycles of succession, and diminishes between-habitat diversity. Given that natural ecosystems respond to natural disturbances, one can expect cyclical processes. Loucks, studying community development in Wisconsin forests, for example, suggested that this development should be viewed as "repeating wave-form phenomena triggered by random perturbations with intervals of 30 to 200 years or more." (These intervals can be longer or shorter in other types of vegetation.) He describes this as a "*stationary process*, with random perturbation (in this case forest fires)."[40] The changes that take place in succession are a characteristic series of transient phenomena and, taken together, they make up a stable system capable of repeating itself with every new perturbation. These elements of the overall stationary process allow for the selection of species which fulfill specialized functions at each of the several phases. This wave phenomenon, then, actually acts as an isolating mechanism and contributes to the evolution and maintenance of diversity. Human control of environments interrupts the natural rhythmic alternation of high productivity phases with high diversity phases. Over time, human management for increased productivity diminishes diversity and eventually undermines the process of natural rejuvenation.[41] The small, incremental value of putting more land into productive monocultures could lead, after decades or centuries, to a complete collapse of that land's productive potential.

A number of ecologists, then, believe that cumulative species extinctions (or perhaps even one extinction of a crucial species) could have cataclysmic effects. This amounts to saying that each decrease in total diversity constitutes a zero-infinity dilemma. Areas seem to have some minimum *threshold* for total diversity; if there are many species extirpations from an area and this threshold is exceeded, an ecosystem breakdown will occur. If species extinctions share Page's nine characteristics, plus the added characteristic of ascending probabilities as more cases occur, then they must be treated as examples of zero-infinity dilemmas of serious magnitude.

It is difficult to assign values to events where the probability of occurrence is low but the probability of irreparable harmful consequences, should they occur, is high. Page notes that the Rasmussen report on risks of catastrophic core meltdown of a nuclear generator sets the probability of such an event at one in five billion per reactor year. But he also notes that the direct historical record based on actual reactor years with no meltdown is equally compatible with a risk assessment of one in one thousand chances per reactor year. Mishan and Page argue that such assessments cannot be given any quantitative validity. Rather, they argue that the issue is whether to adopt one of two rules:

> Rule A would countenance the initiation or continuance of an economic activity until the evidence that it is harmful or risky has been established beyond reasonable doubt. Rule B, in contrast, would debar the economic activity in question until evidence that it is safe has been established beyond reasonable doubt.[42]

They then argue that, far from allowing an assessment of costs and benefits, such choices reduce to questions concerning the state of the society, the desperation of the need for more goods of the type which incur the risk and, ultimately, to what amounts to an attitudinal choice of risk-acceptance or risk-aversion. They conclude, in discussing the analogous case of ozone depletion, that:

> a valid cost-benefit calculation of actions to protect the earth's ozone shield cannot be undertaken in the present state of our ignorance concerning the relevant physical relationships and, therefore, in the present state of our ignorance concerning the nature and magnitude of the risks posed by existing economic activities.[43]

The basic problem here is that the feared danger is so great that *if* there is even a very small likelihood of its occurrence, one should be cautious. But assessment of very low probabilities is difficult because such assessments must be based on a very large track record. The only way to get direct evidence for such assessments is to incur the risk involved over a very large number of cases—a very dangerous thing to do when the disaster is irreversible.

Thus, while the first two arguments establish that each species has some prima facie value, this third argument enhances that value significantly. When the premise that diminutions in diversity create further diminutions is supplemented by the premise that the downward diversity spiral has already begun and is accelerating at an alarming rate, each species takes on an added value. Each species loss carries with it some risk of a catastrophic ecosystem breakdown and increases the risk that the next species loss will result in such a breakdown.

VI. Selective Extinction and Human Utility

Argument 4

In demonstrating the prima facie value of each and every individual species, I have hitherto emphasized characteristics common to every species, regardless of their populational or individual characteristics. I want now to concentrate on characteristics shared by extremely broad groupings of species. If, as scientists agree, species are not equally susceptible to human-caused extinction, it is of interest to know the general characteristics of the most susceptible species.[44] I will argue that those general categories of species which are most susceptible to human-caused extinction are also the most likely to be useful to humans.

Human activities endanger species mainly through habitat destruction and fragmentation, by hunting, by dispersing pollutants through the atmosphere, and by introducing foreign species. Geerat Vermeij has concluded that habitat destruction and fragmentation and hunting are the most important factors in the immediate wave of extinctions and endangerments.[45] Which species are most likely to be threatened by these human activities? There seems broad agreement that the following factors increase likelihood of endangerment or extinction:

(1) rarity (either sparse distribution over a wide range or confinement to a narrow range);[46]
(2) large individual size;[47]
(3) relative height in trophic level;[48]
(4) biotically controlled evolution[49]
(5) low dispersibility, few offspring, and long individual life spans (K-selected species);[50]
(6) specialization of habitat;[51]
(7) involvement in mutualisms and coevolutionary arrangements;[52]
(8) existence in ecosystems of high diversity.[53]

These characteristics are by no means independent of each other. K-selected species invest relatively small amounts of energy in reproduction and tend also to have large body size. They have low dispersibility because they have fewer offspring. But individuals tend to live longer and survive by being superior competitors—size being a comparative advantage in many competitive relationships.[54] These characteristics are, in turn, correlated with another: many K-selected species are also "biotically competent." That is, they evolve largely in response to adaptational pressures emerging from contacts with other organisms, rather than in response to abiotic factors such as climate, soil conditions, etc. Thus, such species are also more involved in mutualisms and coevolutionary arrangements.

They are also likely to occur in specialized habitats characteristic of highly diverse and interconnected ecosystems. In these ecosystems, species at the upper end of relatively long food chains are most susceptible to extinction.[55] The inefficiency of energy transfer from trophic level to trophic level ensures that these upper-level species will be comparatively rare and their dependency on lower levels also increases susceptibility.[56]

Species with these characteristics are also more likely to be useful for human purposes. First, the loss of species from highly interrelated systems is more likely to cause a cascade of further extinctions. Ecosystems of high diversity, and especially complexity of food webs and other inter-relationships, have evolved so as to damp fluctuations caused by minor disturbances. But as they do so, the very interconnections which protect the system against minor fluctuations make it more likely that major disturbances such as a species extinction will have serious reverberations throughout the system.[57] Thus, the very fact that a species is threatened implies that its extinction is more likely than that of a randomly chosen species to contribute to ecosystem deterioration. The very endangerment of a species suggests that its preservation is of special importance in protecting humans from the possibility of ecosystem deterioration and collapse.

Second, the sorts of specialized, biotically competent species described above as more susceptible to extinction are also more "interesting" for human purposes.[58] If each species is viewed as a solution to a group of environmental problems, the biotically competent species have solved the most interesting and complex ones, because they have evolved in response to a complex, diverse, and highly interconnected system. Humans usually find a species of utilitarian value if its members produce some useful substance or if the species has solved some problem that humans also face. Most of the natural substances which have yielded useful medicines evolved as chemical defense mechanisms against predation. The hump-back whale has solved the problem, which humans share, of long-range underwater communication. Current research is designed to understand and perhaps apply those solutions to the human problem.[59]

Generalist species which live short lives and survive by high dispersi-bility are much less in need of such defenses and artifices. These are the pestiferous and weedy species which make human lives less pleasant. Insofar as they have solved environmental problems, they have resorted to a few general and common strategies, and they offer far less fertile ground for research aimed at producing useful goods and services. There-fore, the very species most susceptible to extinction are the ones humans are most likely to find useful, either because of the chemical compounds they produce or because of their ingenious adaptations to complex environmental problems.

Third, since humans are themselves K-selected, i.e., have comparatively large body size, are high in trophic level, exist in highly complex, even social, systems, etc., it is not surprising that the susceptible species under discussion may be studied to shed light on human life and human society. The highly evolved species which share characteristics (1)–(8) listed above are more likely, for example, to be useful in experiments concerning human health. Likewise, the complex social relationships represented within and between species of this sort have far closer analogies to human society than those of species not biotically evolved.

Finally, it is these complex and highly evolved organisms which are aesthetically pleasing to us. They may be pleasing because they share characteristics with humans and consequently excite empathy. Or, they may, like ants and termites, interest us with their extraordinary social adaptations which are so different in organization and mechanism from human societies.[60]

When a biotically competent, specialist species is extirpated from a system, its niche is usually not filled by another specialist species.[61] The specializations of such species are too great to allow interchangeability. Instead, such species are likely to be replaced by generalist, weedy species which colonize quickly and achieve ready dominance. Over long periods of time, if there are no further disturbances, more specialization may occur. But once a system has been sufficiently disturbed to cause an extinction, disturbances reverberate through it. There is great danger that a process of system simplification will have begun. Recovery from serious breakdowns of this sort can take millennia.[62] In the meantime humans will, if they can survive at all, share the planet with comparatively dull, generalist, weedy, and pestiferous species.

VII. TOTAL DIVERSITY AND HUMAN UTILITY

Each species contributes to total diversity in two senses. It constitutes one unit of total diversity, and it also contributes causally to the generation and protection of other species with which it interacts. The final argument for preserving species derives from a list of benefits to humans of living in an area of great total diversity.

Argument 5

Because total diversity is a function of within-habitat diversity and between-habitat diversity, the most diverse areas will contain habitats and ecosystems with very different structures. They will, in other words, be very patterned, patchy environments. At least some of these ecosystems will also be very diverse, with a large number of species packed into highly specialized niches. These very diverse ecosystems will be the comparatively mature systems, as diversity increases during ecological time, especially in the early stages of succession. The most mature systems may, in fact, be somewhat less diverse than those in slightly earlier stages, as the greatest diversity often occurs before climax species have fully exerted their dominance by eliminating all species characteristic of earlier stages. Nevertheless, the existence of some mature communities in an area must contribute to its total diversity, as some species occur only in them.

An area with maximal or near maximal total diversity will also be characterized by intense competition between species as each species attempts to colonize and occupy new areas. And, as habitats become saturated, niches will be partitioned. This partitioning may occur in response to existing abiotic structures such as slopes, rock formations, or soil conditions which encourage species to segregate themselves, with each species occupying only its ideal sites. Differential disturbance also adds to the texture of an area. Limited, noncatastrophic natural and man-made disturbances can create patches and habitats where fugitive species can survive. While a major disturbance decimating an entire area decreases diversity, minor disturbances affecting smaller areas actually contribute to total diversity by restraining climax species from achieving a monoclimax across the whole area and by creating patches of early successional vegetation within more mature areas. Biotic changes, such as the development of closed canopies in some areas, or allelochemical additions to soils, can also produce varying structures which favor particular species in one area while excluding them from niches more suited to competitors in other areas. The result, over ecological and evolutionary time, is an ever greater diversity within habitats which, in turn, contributes to further diversity in habitat structure, creating a self-augmenting spiral of increased competition and diversity.

What, then, are the benefits to humans of living in a diverse area containing many species and varied habitats? First, and most obviously, it can be expected that more useful species will be available in a diverse area. An area with groves of nut trees, blueberry bogs, and raspberry thickets offers useful and delightful food options.

Diverse ecosystems and habitats perform considerable "ecosystem services."[63] Vegetation provides oxygen and moderates climate and hydrological cycles. Marshes and bogs provide tertiary treatment of human wastes. Undisturbed, diverse ecosystems harbor birds and spiders which aid in the control of insect pests. The list can be augmented almost indefinitely. And nobody can deny the aesthetic benefits of a diverse and multi-textured environment. The human spirit seems to crave diversity and natural settings. A diverse area offers a variety of such settings and improves the quality of human life.

Residents of areas of great diversity reap other long-term advantages. Diversity contributes to the "regenerative cycle" of nature. A patchy, highly textured environment, containing some recently disturbed areas in early successional stages and some more fully mature areas, is able to regenerate nutrients, especially if the disturbances take place in a shifting pattern. In order to explain this point I will contrast an agricultural monoculture with a more natural, evolved, and highly organized system. Enforced monocultures can be extremely productive, and almost every form of modern agriculture involves methods of cultivation and nutrient addition which hold the community in early stages of succession. Species important as agricultural crops are opportunistic species that turn their energy into reproduction, wasting little on building structure in the form of woody stems, etc. Modern agriculture maintains its productivity by holding back successional development and by exploiting the energy spent in reproduction in the form of fruit, grain, etc. of opportunistic species. But such high productivity is costly in that harvesting removes energy from the system. Even more crucial, however, is the fact that cultivation has maintained the system at early levels of succession and the food web is not closed. Little primary production finds its way back into the store of nutrients available. Most highly cultivated monocultures may be maintained in their productive state only through the importation of great amounts of energy in two separate forms. First, they require energy (work) to maintain them at the early successional stage—today, this energy comes increasingly from fossil fuels. Second, they require massive inputs of nutrients (fertilizer) to maintain production as cropping removes energy and an open nutrient cycle leads eventually to nutrient impoverishment.

A "patchy" area containing a variety of ecosystems at differing stages of development will have greater total diversity. When land is no longer productive in monocultures, it may lie fallow and, if species are available to develop a productive and closed ecosystem, nutrients will be stored in the soil and it will become productive once again. But if only opportunistic species remain, such development will be slow, as in abandoned areas in the centers of large cities. A high level of diversity, then, provides extremely valuable insurance against downward spirals in productivity. An ideally productive environment would intersperse patches of monoculture with natural systems. If these uses are rotated, species will move from one area to another. They can, therefore, contribute to the rejuvenation of areas where intense use leads to deterioration. Such a situation has some hope of providing sustained productivity without overtaxing nonrenewable resources such as chemical nutrients and fossil fuels.

Since most monocultures are highly productive only at great cost to nonrenewable resources, they depend upon energy from fossil fuels and nutrients created aeons ago by the very systems they are increasingly replacing. Using up nonrenewable resources at unnecessarily rapid rates is, essentially, spending capital. Those resources will increase in value in the future and using them now to maximize production decreases the future's wealth.

VIII. Conclusion

I began by contrasting two ways of framing the question, what is the value of a species? It has been common to approach this question by isolating a particular species, examining it for useful individual or populational characteristics, and placing a dollar value on those uses. If no such uses are found, the value assigned would be the product of the value it would have if it turned out to be useful and the probability that such a use might be found in the future. This approach is associated with the modified benefit-cost analysis (BCA) approach advocated by the Resources for the Future group, as represented by Anthony Fisher and others.[64] It has two essential elements: it examines species individually, and it assigns quantified dollar values to species which can be factored into analytic decision procedures to guide decision-makers in cases where steps to protect a species incur costs. Although Fisher gives the benefit of the doubt to the species when the costs of preservation are roughly in balance with the value of a species computed in this way, this approach amounts to requiring that the preservation of each species be accomplished piecemeal. Every species must exhibit some actual or potential value to justify its existence.

I have argued that this modified BCA approach will systematically undervalue species, by providing five arguments for believing that *every* species has utilitarian value. On my approach, the question becomes one of ascertaining the value of total diversity and of any species as a unit of total diversity.

My argument can be seen as operating in stages. The first stage, represented by the first two arguments, establishes the contributory value of species. Every species saved provides opportunities for new species to emerge and, consequently, serves humans indirectly by supporting other species which will prove to be valuable.

The third argument supplements the initial, prima facie value of each species by adding the premise that we have already embarked upon an alarming downward spiral in the total diversity of the earth. Such a downward spiral raises the specter of a major ecosystem collapse, and this implies that each extinction must be viewed as a zero-infinity dilemma—each extinction has a small probability of leading to disastrous consequences. No scientist doubts that, at some point on that difficult-to-reverse spiral, humans will suffer great losses of utility—perhaps even their own extinction.

The fourth argument exploits generalizations about the types of species susceptible to extinction. It turns out that, for several reasons, the more extinction-prone species are precisely those which are most likely to be useful to humans. Thus, the utilitarian benefit to humans of any species, as shown in the first three arguments, is multiplied if the species is endangered. The very fact that it is endangered signifies an increased likelihood that its loss will adversely affect human utilities.

Finally, humans derive a whole range of values—economic, commercial, and aesthetic—from living in an area of great total diversity. One such value is the regenerative value of complex and mature ecosystems. Only complex, mature systems can produce and store the resources necessary to regenerate land abused by monocultural overexploitation.

These arguments provide sufficient reason to reject the BCA approach to valuing species. That approach systematically undervalues species by concentrating on commercial and economic values manifest in characteristics of particular species. It thus ignores the considerable contributory values, regenerative values, and protectionist values shared by all species, viewed as units of total diversity. These values are not quantifiable—they cannot be assigned dollar values. Thus, the BCA approach embodies a false assumption—that it is possible to assign meaningful dollar values to species. It is pointless, therefore, to apply BCA approaches to decision-making regarding endangered species.

Do these arguments imply that economic arguments could never justify extinguishing a species? It is possible that a project's economic benefits might be so great and so essential that these considerations might be overridden. But the implication is that these economic gains must be seen as great enough to justify the destruction of an irreplaceable resource for the future, a resource whose loss will affect the earth's total diversity. If the gains are sufficiently great, some of them should be transferred to funding the protection of other species and natural areas.

The arguments of this paper support and strengthen a different economic criterion for deciding issues involving endangered species, the safe minimum standard of conservation (SMS) approach of S. V. Ciriacy-Wantrup, Richard Bishop, and Oliver Ray Stanton.[65] That approach says that the safe minimum standard of conservation should be adopted.

According to the SMS criterion, an irreplaceable resource such as a species should be saved unless the social costs are "unacceptably large." This standard therefore assumes that any species is valuable and that it should be saved unless some extraordinary costs would be incurred in its protection. The SMS approach mainly ignores benefit computations, assuming that each species has considerable value. This chapter supports the SMS approach by providing ecological theory and data which explain why its central assumption (that every species is prima facie valuable) is true. It also gives some substance and precision to the vague rule that a species should be saved unless the social costs are unacceptably large. Ideally, such a rule would be supplemented with an operational method for assigning a dollar value to species in general and for adjusting that value, depending upon particular characteristics of the species involved. My five arguments entail that no such method is actually possible. But they do detail what is at stake in species preservation and thereby give more substance to the SMS rule. I believe they show that each species should be accorded substantial value and that, when a species has particular, identified uses, the values derived from those uses should be added to that original, general value. Even though no specific dollar amount can be assigned as the value of a species, a rational and forward-looking society would adopt the preservation of species as a general policy, rather than as an open question to be decided on a case-by-case basis. This is not to say that the policy could never be violated, but violations should be advocated only under extraordinary conditions.

A concluding analogy may be useful. Consider a confirmed alcoholic deciding whether to have a drink after learning that he has suffered irreversible liver damage. Such a decision involves a trade-off between the advantages of having the drink and the undoubtedly minimal liver damage to be expected from one more drink. So conceptualized, the decision will be made in different ways in different situations, depending upon the perceived benefits of taking the drink. But medical experts, knowing that it is virtually impossible for a confirmed alcoholic to have just one drink, counsel that the alcoholic adopt a policy against taking even a single drink. To treat particular decisions as to whether to have a drink as a series of incremental decisions prejudices the case in a disastrous manner.[66]

My point is similar. The scientific evidence that diminutions in diversity compound themselves through time is analogous to the medical evidence that alcoholics cannot take just one drink. To treat the decision to allow a species to go extinct as an isolated assessment of the value of a single species versus some short-term economic gain misconstrues the issue. A forward-looking society should have a policy against decreasing biological diversity, as the alcoholic should have a policy against taking any drink at all. Both policies could be overridden. No doctor would tell an alcoholic to die of thirst when no nonalcoholic liquids were available. But the doctor would surely recommend that the alcoholic avoid such difficult situations at considerable cost and that truly extraordinary conditions must obtain before taking a drink becomes rational. These arguments imply that, on the SMS approach, almost all costs of preserving a species should be considered "reasonable," thereby giving substance to and strengthening the SMS criterion.

NOTES

1. For example, a symposium on "Estimating the Value of Endangered Species: Responsibilities and Role of the Scientific Community" was held at the American Association for the Advancement of Science Meetings, Washington, D.C., January 4, 1982. Also see Randall's chapter, this volume.
2. David Ehrenfeld, *The Arrogance of Humanism* (New York: Oxford University Press, 1981), p. 177.
3. See chapters by Donald Regan and J. Baird Callicott, this volume.
4. See Aldo Leopold, *A Sand County Almanac* (New York: Oxford University Press, 1975), p. 210.
5. Stuart H. Hurlbert, "The Nonconcept of Species Diversity: A Critique and Alternative Parameters," *Ecology* 52 (1971): 577-86.
6. See R. H. Whittaker, "Vegetation of the Siskiyou Mountains, Oregon and California," *Ecological Monographs* 30, no. 3 (July 1960): 320; and Robert H. MacArthur, "Patterns of Species Diversity," *Biological Review* 40 (1965): 510-33.
7. R. H. Whittaker, "Evolution and Measurement of Species Diversity," *Taxon* 21 (1972): 232.
8. Ibid.
9. Simon Levin, "Population Dynamics and Heterogeneous Environments," *Annual Review of Ecology and Systematics* 7 (1976): 287-310, esp. 302-303. Also see Peter S. White, "Pattern, Process and Natural Disturbance in Vegetation," *The Botanical Review* 45 (1979): 277.
10. C. S. Holling, "Resilience and Stability of Ecological Systems," *Annual Review of Ecology and Systematics* 4 (1973): 1-23.
11. See Whittaker, "Species Diversity," p. 232.
12. Aldo Leopold, "Conservation," in *Round River: From the Journals of Aldo Leopold*, edited by Luna B. Leopold (New York: Oxford University Press, 1953), p. 147.
13. Charles S. Elton, *The Ecology of Invasions by Plants and Animals* (London: Methuen, 1958), pp. 54ff.; also see Gerald Lieberman, "The Preservation of Ecological Diversity: A Necessity or a Luxury?" *Naturalist* 26 (1975): 24-31; J. W. Humke et al., *Final Report. The Preservation of Natural Diversity:*

A Survey and Recommendations, Report prepared for the U.S. Department of the Interior by the Nature Conservancy, 1975; Gerald O. Barney, *The Global 2000 Report to the President: Entering the Twenty-First Century* (New York: Penguin Books, 1980).

14. See, for example, William H. Drury and Ian C.T. Nisbet, "Succession," *Journal of the Arnold Arboretum* 54 (1973): 331-68.

15. For classic accounts of patchiness in succession, see A. S. Watt, "Pattern and Process in the Plant Community," *Journal of Ecology* 35 (1947): 1-22, and R. H. Whittaker, "A Consideration of Climax Theory: The Climax as Population and Pattern," *Ecological Monographs* 23 (1953). Comprehensive recent accounts include John L. Harper, *Population Biology of Plants* (London: Academic Press, 1977), chap. 12; Peter S. White, "Pattern, Process and Natural Disturbance in Vegetation," *The Botanical Review* 45 (1979): 229-97; Simon Levin, "Dispersion and Population Interactions," *The American Naturalist* 108 (1974): 207-28; and Levin, "Population Dynamic Models in Heterogeneous Environments," *Annual Review of Ecology and Systematics* 7 (1976): 287-310. See also Drury and Nisbet's positive account of spatial and temporal sequences of vegetation patterns.

16. Harper, *Population Biology*, p. 711; E. C. Pielou, "Species Diversity and Pattern—Diversity in the Study of Ecological Succession," *Journal of Theoretical Biology* 10 (1966): 370-83.

17. Levin, "Population Dynamics," p. 294.

18. Martin L. Cody, "Towards a Theory of Continental Species Diversity," in *Ecology and Evolution of Communities*, edited by Martin L. Cody and Jared M. Diamond (Cambridge, Mass.: Harvard University Press, 1975).

19. Ibid., p. 244. This account squares with that of Robert H. MacArthur and Edward O. Wilson, *The Theory of Island Biogeography* (Princeton, N.J.: Princeton University Press, 1967), pp. 23, 171, 176; also see Robert K. Peet, "Ecosystem Convergence," *The American Naturalist* 112 (1978): 441-44.

20. R. H. Whittaker, *Communities and Ecosystems* (New York: Macmillan, 1970), p. 103.

21. Peter Raven, Testimony before the Senate Subcommittee on Environmental Pollution, Committee on Environment and Public Works, United States Senate. Ninety-Seventh Congress, First Session, December 8 and 10, 1981. Printed in *Endangered Species Act Oversight* (Washington, D.C.: U.S. Government Printing Office, 1982), p. 293. Also see John Terborgh and Blair Winter, "Some Causes of Extinction," in *Conservation Biology: An Evolutionary Ecological Perspective*, edited by M. E. Soule and B. A. Wilcox (Sunderland, Mass.: Sinauer, 1980), p. 89.

22. Charles W. Fowler and James A. MacMahon, "Selective Extinction and Speciation: Their Influence on the Structure and Functioning of Communities and Ecosystems," *The American Naturalist* 199 (1982): 483; Terborgh and Winter, "Extinction."

23. For example, the most recent theories describing events occurring at the time of mass extinctions at the end of the Cretaceous Era suggest that a single

major event (speculation suggests an asteroid or disintegrated comet striking the earth) led to a cycle of extinctions which occurred over a period of 50,000 years. Kenneth J. Hsu et al., "Mass Mortality and Its Environmental and Evolutionary Consequences," *Science* 216 (1982): 249-50.

24. This objection was suggested to me by Henry Shue and Robert Fullinwider.

25. See, for example, William K. Frankena, *Ethics* (Englewood Cliffs, N.J.: Prentice-Hall, 1963), p. 36.

26. Paul and Anne Ehrlich, *Extinction: The Causes and Consequences of the Disappearance of Species* (New York: Random House, 1981), p. 7.

27. Norman Myers, *The Sinking Ark* (Oxford: Pergamon Press, 1979), p. 4.

28. Ibid.

29. Ibid. Myers acknowledges that these figures are artificially low, as they are limited almost entirely to birds and mammals and do not count unknown and unnamed species lost (p. 31).

30. Ibid.

31. Ibid., p. 5.

32. Ibid.

33. Ibid.

34. See Stephen Stanley, *Macroevolution: Pattern and Process* (San Francisco: Freeman, 1979), pp. 106-107, 135-36.

35. Myers, *The Sinking Ark*, p. 17.

36. Eugene Morton, "The Realities of Reintroducing Species into the Wild: The Problem of Original Habitat Alteration," Lecture at First Annual Zoological Park Symposium, Washington, D.C., September 11-12, 1982.

37. Ehrlich and Ehrlich, *Extinction*, pp. xi-xvi.

38. Talbot Page, "Keeping Score: An Actuarial Approach to Zero-Infinity Dilemmas," Social Sciences Working Paper no. 248 (Pasadena, Calif.: Division of Humanities and Social Sciences, California Institute of Technology, January 1979).

39. Talbot Page, "A Generic View of Toxic Chemicals and Similar Risks," *Ecology Law Quarterly* 7, no. 2 (1978): 207-44.

40. Orie Loucks, "Evolution of Diversity, Efficiency, and Community Stability," *American Zoologist* 10 (1970): 23-24.

41. Ibid., p. 25.

42. Ezra Mishan and Talbot Page, "The Methodology of Cost Benefit Analysis with Particular Reference to the Ozone Problem," Social Science Working Paper no. 249 (Pasadena, Calif.: Division of Humanities and Social Sciences, California Institute of Technology, January 1979), p. 71.

43. Ibid., p. 75.

44. E. O. Willis, "Populations and Local Extinctions of Birds on Barro Colorado Island, Panama," *Ecological Monographs* 44 (1974): 153-69; Terborgh and Winter, "Extinction"; Robert E. Jenkins, "Endangerable Species," *Ecology Forum* 25; and Lawrence Slobodkin, this volume.

45. Geerat J. Vermeij, this volume.

46. Terborgh and Winter, "Extinction," p. 128; Jenkins, "Endangerable Spe-

cies," p. 21; Fowler and MacMahon, "Selective Extinction," p. 482; and Slobodkin, this volume.

47. Jenkins, "Endangerable Species," p. 21. See also Paul A. Colinvaux, *Why Big Fierce Animals Are Rare* (Princeton, N.J.: Princeton University Press, 1978).

48. Ibid., p. 21; Fowler and MacMahon, "Selective Extinction," p. 483.

49. Geerat J. Vermeij, *Biogeography and Adaptation* (Cambridge, Mass.: Harvard University Press, 1978), p. 183; and Vermeij, this volume.

50. MacArthur and Wilson, "Island Biogeography," p. 151; Vermeij, *Biogeography*, pp. 186ff.

51. Terborgh and Winter, "Extinction," p. 21.

52. Fowler and MacMahon, "Selective Extinction," p. 483.

53. Gordon H. Orians, "Diversity, Stability and Maturity in Natural Ecosystems," in *Unifying Concepts in Ecology*, edited by W. H. Van Dobben and R. H. Lowe-McConnell (The Hague: Junk, 1975), pp. 139-49.

54. J. P. Grime, "Evidence for the Existence of Three Primary Strategies in Plants and Its Relevance to Ecological and Evolutionary Theory," *The American Naturalist* 111 (1977): 1189; Fowler and MacMahon, "Selective Extinction," p. 485.

55. Fowler and MacMahon, "Selective Extinction," p. 483.

56. Lawrence Slobodkin, "Ecological Energy Relationships at the Population Level," *The American Naturalist* 95 (1960): 213-36.

57. Raven, "Testimony," p. 293; Fowler and MacMahon, "Selective Speciation and Extinction"; Terborgh and Winter, "Extinction," p. 133; D. Futuyma, "Community Structure and Stability in Constant Environments," *American Naturalist* 107 (1973): 443-46.

58. Vermeij, this volume.

59. Terry Leitzell, "Extinction, Evolution, and Environmental Management," this volume.

60. Ehrlich and Ehrlich, *Extinction*, chap. 3.

61. Hsu et al., "Mass Mortality," p. 255; Peet, "Ecosystem Convergence," p. 442.

62. Vermeij, this volume.

63. See Ehrlich and Ehrlich, *Extinction*, chap. 5; Eugene Odum, "The Strategy of Ecosystem Development," *Science* 164 (1969): 266.

64. Anthony C. Fisher, "Economic Analyses and the Extinction of Species," Energy and Resources Group, University of California, Berkeley (Report No. ERG-WP.81-4); V. K. Smith and J. V. Krutilla, "Endangered Species, Irreversibility, and Uncertainty: A Comment," *American Journal of Agricultural Economics* 61 (1979): 371-75; J. R. Miller and F. C. Menz, "Some Economic Considerations in Wildlife Preservation," *Southern Economic Journal* 45 (1979): 718-29: Charles Ploudre, "Conservation of Extinguishable Species," *Natural Resources Journal* 15 (1975): 791-98; Alan Randall, this volume.

65. S. V. Ciriacy-Wantrup, *Resource Conservation: Economics and Politics* (Berkeley and Los Angeles: University of California, Division of Agricultural

Sciences, 1968); Richard Bishop, "Endangered Species Uncertainty: The Economics of a Safe Minimum Standard," *American Journal of Agricultural Economics* 60 (1978): 10-18; Richard Bishop, "Endangered Species: An Economic Perspective," *Forty-Fifth North American Wildlife and Natural Resources Conferences Transactions* (1980): 208-18; Oliver Ray Stanton, untitled remarks delivered at AAAS meetings; also see Randall, this volume.

66. See Thomas E. Lovejoy, this volume.

6

On the Intrinsic Value of
Nonhuman Species

J. BAIRD CALLICOTT

I. The "Facts" and the Values

At present the earth is in the throes of an episode of biotic impoverishment of, perhaps, unprecedented magnitude.[1] The current rate of species extinction is the subject of controversy, but all parties agree that it is alarmingly great, and accelerating.[2] From 1600 to 1900 the average rate of species extinction was roughly one every four years, from 1900 to the present one per year;[3] and, according to Norman Myers, "if present average patterns of exploitation persist," the rate of extinction during the last quarter of the twentieth century may reach something over 100 species per day![4] It is conceivable that by the end of the century one million or more species could become extinct.[5] Already gone are, in the words of Alfred Russel Wallace, "the hugest, and fiercest, and strangest forms."[6] Next to go will be myriad species of plants and invertebrates.

Well, so what? Why should we care? Aren't more than 90 percent of the species ever to have existed on earth now extinct?[7] Isn't species extinction, after all, a natural process?

Undoubtedly, species extinction is natural, in the sense that all natural phenomena are natural. (Certainly it is not supernatural.) Species extinction, when compensated by speciation, moreover, is normal as well as natural. But massive, abrupt species extinction and consequent biological impoverishment are *not* normal. The fossil record indicates that several discernible mass extinction events occurred in the geologic past, but it also indicates that the rate of "background" or routine extinctions has declined with time and, correspondingly, that biological diversity has increased with time.[8] On the average, speciation has outpaced extinction. The earth's evolutionary process tends toward greater biological diversity (and I do not mean to suggest here anything teleological), although it has been interrupted by widely spaced "setbacks."

To know that massive species extinction is abnormal, albeit natural, and that the tendency of organic evolution is toward greater biological diversity, however, does not settle the question of value. Why should one species be concerned about the threat of destruction it poses to others? More to the point, why should we, *Homo sapiens*, preserve and nurture those species yet surviving?

Many cogent arguments for species preservation of a kind vulgarly called "utilitarian" have recently appeared.[9] Any argument for species preservation which is addressed to human welfare or human happiness (whether material or spiritual) is essentially "utilitarian" in the received sense of the term.[10] There are, in general, two kinds of value: (1) intrinsic value and (2) instrumental value. A "utilitarian" or "homocentric" argument for species preservation either explicitly or implicitly assumes that human beings (or, more abstractly, human welfare or happiness) are intrinsically valuable and that all other things, including other forms of life, are valuable only as means or instruments which may serve human beings (or facilitate human welfare or happiness).

One often finds, however, lurking beneath a recitation of the benefits to man provided by other species severally and the existing biotic complement of earth collectively a scarcely concealed nonutilitarian substratum of value. George Woodwell, for example, wistfully remarks that "one might dream that on the only green planet we know, life would have a special value of its own." But, since such a value is not universally acknowledged, he and his colleague Howard Irwin agree that it is necessary to argue for species preservation in terms "understandable and usable in politics," and which "the public can easily understand and accept," i.e., utilitarian terms.[11] The utilitarian arguments, in other words, seem often to be a way of selling the public on policies that are felt to be somehow right independently of present and future human well-being. One suspects that for Woodwell, Irwin, and many other ardent advocates of species preservation the conventional utilitarian arguments are but a subterfuge and that their deeper concerns emanate from other ideals and values.[12] Many distinguished conservationists, indeed, have openly and boldly declared that other species have a right to exist or conversely that we, *Homo sapiens*, have no right to cause their extinction.[13]

While the utilitarian case for species preservation has been fully and persuasively articulated, the nonutilitarian case has been neglected. Upon consideration of the philosophical problem of providing a nonutilitarian case for species preservation Alastair Gunn has recently even expressed despair: "It seems impossible to provide *reasons* for valuing natural kinds.

It seems to me that the world would be a worse place if we were to lose the tiger, the bald eagle, or the various species of whale, but I do not know how to justify this view to someone who disagrees."[14] In the absence of well-considered reasons, it is simply asserted that we *Homo sapiens* have "moral obligations" in regard to other species or that other species have a "right to exist" or that we have "no right to render them extinct" and left at that. In this chapter I shall try to provide the missing discussion of the "rights" (the intrinsic value) of other species. First, I shall critically explore the concept of "species rights." I shall then turn to my principal task: a discussion of several distinct axiologies which may provide intrinsic value for nonhuman forms of life. While arguments for species preservation based upon the aesthetic value of other species may sometimes seem genuinely disinterested or biocentric, they are, nevertheless, readily reducible to a homocentric or utilitarian form: other species, in the final analysis, are valuable as aesthetic resources for the aesthetic enjoyment they afford (some) people.[15] Appeals for species preservation based upon the alleged rights of other species are more resistant to reduction.

An analysis of the concept of rights, sufficient for the needs of this discussion, is provided in the next section. There I argue that the assertion of rights on behalf of species, taken literally, is incoherent and thus that the persistent popular call for "species rights" is essentially symbolic. What it symbolizes, i.e., what it imprecisely but dramatically expresses, I suggest, is the widely shared intuition that nonhuman species possess intrinsic value. Since this concept, the concept of "intrinsic value," is, though traditional, somewhat technical and absolutely central to the main body of this discussion, it should, perhaps, be explicitly defined here at the outset.

Something is intrinsically valuable if it is valuable *in* and *for* itself—if its value is not derived from its utility, but is independent of any use or function it may have in relation to something or someone else. In classical philosophical terminology, an intrinsically valuable entity is said to be an "end-in-itself," not just a "means" to another's ends.

Most systems of modern ethics, both formal philosophical systems (e.g., Kant's deontology), and less formal popular systems (e.g., the Christian ethic), take it for granted that human beings are intrinsically valuable, that, in other words, each human being is valuable in and for himself/herself independently of any contribution s/he may make to the welfare of another person or to society collectively. We may not discard or destroy worn-out, broken, or imperfectly made human beings as we might tools in similar condition because human beings are, it is almost universally supposed, intrinsically—not, like tools or "resources," merely instrumentally—valuable.

Accounts of intrinsic value in the Western philosophical tradition have varied considerably. Plato, who did not share the modern dogma that human beings are paradigmatic cases of intrinsically valuable entities, posited a "form," the Good, as the ultimate source of intrinsic value for intrinsically valuable entities. I shall have occasion in a subsequent section of this chapter to explain Plato's understanding of the Good, i.e., of intrinsic value, more fully, and apply it to my central theme, the problematic intrinsic value of nonhuman species. Aristotle, somewhat less abstractly and elusively than Plato, concluded that happiness is the only intrinsically valuable thing, among other reasons because, he thought, happiness is the only thing pursued for the sake of itself. Kant, to whom I also refer more fully below, rested the intrinsic value of persons (human beings) on our capacity to reason. G. E. Moore, whose account is not outlined here and applied to the problem of the value of species because in the final analysis it appeals to mute intuition, thought that intrinsic value was a primitive non-natural property of objects as the color red is a primitive natural property. One either perceived it, he thought, or one did not. Such a theory as Moore's leaves no room for rational discussion of controversial cases. I may perceive the intrinsic value of species and you may not. Since intrinsic value, so construed, is a primitive non-natural property, I cannot explain why species are intrinsically valuable; I can only accuse you of a kind of moral insensitivity. Moore's theory reduces moral debate to question-begging and/or brow-beating.

A fundamental doctrine of modern science remains a formidable obstacle, however, to all the heroic attempts of philosophers to establish the existence, and adequately explain the nature, of intrinsic value, the value of something in and for itself. The objective physical world is sharply distinguished from subjective consciousness in the metaphysical posture of modern science as originally formulated by Descartes. Thought, feeling, sensation, and value have ever since been, from the point of view of Scientific Naturalism, regarded as confined to the subjective realm of consciousness. The objective, physical world is therefore value-free from a scientific point of view.

Quantum theory, relativity, and the other revolutionary developments of post-modern science are said to have invalidated the Cartesian distinction between the subjective and the objective domains, and hence to promise profound consequences not only for epistemology and ontology, but for value theory as well.[16] The axiological consequences of post-modern science, however, remain at this point programmatic; they have not been worked out in any detail, and seem, in any case, remote and metaphorical. Further, in the structure of science itself, quantum theory has little direct relationship to or influence on biology. Hence, at the level of organization with which we are concerned, the macroscopic world of terrestrial life and the value of its component species, the classical attitude that nature is value-neutral remains a virtually unchallenged dogma of the scientific world view. From this perspective, the attribution of intrinsic value to species, as to anything else under the sun, is doomed at the outset to failure.

On the other hand, many people, including some scientists, persist intuitively to feel that nonhuman species are valuable in and for themselves, quite apart from their usefulness to us as resources, either material or spiritual, or as providers of (human) life-supporting services. In the discussion which follows I explore several possible and preferred grounds for this very genuine and sincere ethical intuition.

The classical scientific world view is not, after all, the only world view represented in Western civilization today. The intrinsic value of species may be quite straightforwardly defended in terms of some elements of the pre-scientific, but still well-represented, Judeo-Christian world view. In the last section of this chapter, I attempt to find a compromise, recommending a theory of "intrinsic value" which at once respects the institutionalized cleavage between object and subject, fact and value of the scientific world view, and yet does justice to the intuition that some natural "entities," nonhuman species among them, are more than merely instrumentally valuable. In the process, the concept of intrinsic value is transformed, or more precisely, truncated.

I concede that, from the point of view of Scientific Naturalism, the *source* of all value is human consciousness, but it by no means follows that the *locus* of all value is consciousness itself or a mode of consciousness like reason, pleasure, or knowledge. In other words, something may be valuable only because someone values it, but it may also be valued for itself, not for the sake of any subjective experience (pleasure, knowledge, aesthetic satisfaction, etc.) it may afford the valuer. Value may be subjective and affective, but it is intentional, not self-referential. For example, a newborn infant is valuable to its parents for its own sake as well as for the joy or any other experience it may afford them. In and of itself an infant child is as value-neutral as a stone or a hydrogen atom, considered in strict accordance with the subject-object/fact-value dichotomy of modern science. Yet we still may wish to say that a newborn infant is "intrinsically valuable" (even though its value depends, in the last analysis, on human consciousness) in order to distinguish the *non-instrumental* value it has for its parents, relatives, and the human community generally from its actual or potential instrumental value—the

pleasure it gives its parents, the pride it affords its relatives, the contribution it may make to society, etc. In so doing, however, "intrinsic value"
retains only half its traditional meaning. An intrinsically valuable thing
on this reading is valuable *for* its own sake, *for* itself, but it is not valuable
in itself, i.e., completely independently of any consciousness, since no
value can in principle, from the point of view of classical normal science,
be altogether independent of a valuing consciousness. Nonhuman species,
I argue, may possess intrinsic value in this truncated sense, which is
consistent with the world view of Scientific Naturalism. Indeed, my suggestion is that the world view of modern science not only *allows* for the
intrinsic value of nonhuman species in this limited sense, but its cos-
mological, evolutionary, and ecological perspectives actually *foster* such
value.

II. "SPECIES RIGHTS" OR SPECIOUS RIGHTS?

There is, I think, a certain aura of mystery surrounding so-called natural
or moral rights (as opposed to civil or legal rights which may be defined
in a positive or operational way).[17] Because it is a noun, "right" seems
to be the name of an entity of some sort. A person possesses shoes, teeth,
kidneys, feelings, thoughts, and certain inalienable natural rights. Feelings
and thoughts may not be entities on a par with kidneys, teeth, and shoes,
but they are at least palpable states of a human organism. Rights are not
even entities of this tenuous sort. The term "rights" is, rather, an ex-
pressive locution masquerading as a substantive.[18] Of course, that is a
big part of its talismanic power.

In this connection it is instructive to note that the concept of a moral
right is both modern and Western. Plato and Aristotle never so much as
mention rights in their ethical and political philosophies. The Bible contains no myth, allegory, sermon, or homily on rights. Oriental religious
philosophies contain little that can be interpreted as pertaining to rights.
Indeed, there appear to be no clear instances of the systematic assertion
and theoretical defense of moral rights before the seventeenth century.
If rights were real natural entities associated with us from birth, it is
surprising that they were not sooner and more universally noticed.

Talk about the "rights" of species to a share of life seems clearly to
be an extension of this relatively recent tradition of Western ethical discourse about natural or moral rights. Having exorcised the ghostly presence of some occult entity that the substantive term "rights" conjures, I
suggest that a fundamental part of its function in popular discourse is
to assert "standing" for someone (or something) in the moral community,
i.e., status as an end rather than a mere means to be used for another's

betterment. If this minimalistic interpretation is correct, then the argument that we ought not to cause the extinction of other species because they have a right to exist not only resists reduction to a utilitarian form, but avows that whatever instrumental value species may (or may not) have, they, no less than we, have intrinsic value too.[19]

It is understandable but regrettable that the moral intuition that non-human species have intrinsic as well as instrumental value is popularly expressed in terms of rights, "species rights." It is understandable because talk of rights has become the usual and preferred way to express moral considerability; but it is regrettable because the concept of species rights taken at face value seems to be philosophical nonsense. The "grammar" of the term "rights" appears to require that those possessing them be, if not persons, at least localizable things of some sort. But the term "species" traditionally designates a class or kind. A class, by definition, is not an individual or localizable thing. How then could it possibly have rights? The proposition itself seems, upon its face, conceptually odd if not logically contradictory.

There are several ways of circumventing this difficulty, but none of them is satisfactory. One would be to follow Plato and hypostatize classes. It would then be logically possible to endow species with rights, although to do so would be pointless since according to the same ontological theory species are eternal Forms and could not therefore be threatened or endangered.

One could argue that talk about "species rights" is just a loose and imprecise way of talking about the putative rights of individual nonhuman organisms. Analogously with "gay rights" or "minority rights," which we understand to devolve upon members of certain classes, "species rights" could be construed to refer not to species per se but to specimens. Such a reduction from type to token, however, would miss the point. Those who claim that nonhuman species have a right to exist are concerned with species preservation, not necessarily with animal and/or plant welfare, an entirely separate issue. It is logically consistent to hold that species have intrinsic value, but that specimens of some species do not. Indeed, I am inclined to think that for some ardent species preservationists, species have intrinsic value while specimens have only instrumental value—as means to the preservation of species. An individual whooping crane, for example, is no more or less valuable than a sandhill crane qua individual, but because a fertile whooping crane carries a significant fraction of the genetic material of her species, her life is a precious instrument for the salvation of her kind. If whooping cranes are ever rescued from the brink, specimens may even be routinely "culled" from the "herd" to improve the "stock."

Finally, one may adopt a nontraditional interpretation of the term "species." David Hull has argued, for example, that the traditional interpretation of "species" as a class designator is theoretically useless in evolutionary biology. In his view, species are "superorganismic entities" or "historical entities," localizable (as classes or kinds are not) in space, however diffusely, and in time, however protractively.[20] Although paradigmatic holders of rights are individual persons, it is not at all unusual or conceptually odd to ascribe rights to "superorganismic entities"— corporations, for example. Nations, to take another example, have certain rights which are not the same as the sum of the rights of their individual citizens. A nation's right to sovereignty is hardly the sum of the respective rights of its several citizens to sovereignty (whatever that may mean). "Species rights" might be understood, in short, by analogy with "national rights." This is the most attractive way to reify the reference of the term "species" so that the phrase "species rights," taken at face value, is at least intelligible. However, Hull's proposed reference for the term "species" has not been universally accepted among philosophers of science.[21] In any case, the assertion of species rights is primarily symbolic; it seems to be less a literal assertion of rights than an assertion in familiar, strong moral terms of the intrinsic value of nonhuman species.

Species rights is indeed a specious notion. It would be better if the notion simply went away. But of course it won't because it expresses in a particularly current and forceful manner of speech a deeply felt and widely shared intuition that species are intrinsically valuable.

Of course, this analysis of certain claims being made on behalf of nonhuman forms of life takes us, so to speak, out of the frying pan into the fire. "Intrinsic value" is no less mysterious a notion than "natural rights." Indeed it is frankly metaphysical. But that is, in fact, its virtue. We do not require a more liberal theory of rights; we need to discover, rather, metaphysical foundations for the intrinsic value of other species which the assertion of rights on their behalf expresses. What are the ethical systems and, more generally, the world views in which claims of the intrinsic value of nonhuman species are embedded? I shall sketch several alternative metaphysics of morals in which the intrinsic value of other species may be grounded.[22]

III. J-Theism

In a deservedly famous discussion of the relative merits of utilitarian and nonutilitarian arguments for species preservation, David Ehrenfeld places the "non-economic [i.e., intrinsic] value" and "unimpeachable right to continued existence" of ecosystems and species in a religious context.[23]

Ehrenfeld invokes the Judeo-Christian religious belief system (to which the concept of natural rights has been grafted since the seventeenth century) as a supporting matrix of ideas for "species rights." He suggests we call it the " 'Noah Principle' after the person who was one of the first to put it into practice."[24]

At first glance the Judeo-Christian world view would seem inhospitable to the suggestion that nonhuman species are intrinsically valuable. "Academic" Christian theology, from Origen and Augustine to Bultmann and Teilhard, has been consistently hostile to the idea that human beings have any duties directly to individual animals and plants (to say nothing of species), precisely because animals and plants lack the requisite qualifications (an immortal soul, the *imago Dei*, or whatever) for membership in the moral community. Orthodox Christian theology, historically, lines up overwhelmingly against the notion that nonhuman creatures considered individually or collectively have any sort of value other than instrumental value or any other role in creation other than to serve man.[25]

Lynn White, Jr., in his celebrated environmentalist critique of the Judeo-Christian world view, traces this attitude to those verses in Genesis (1:26-30) which set man apart from the rest of creation and appear to deliver the creation into his hands.[26] White emphatically declares, indeed, that "Christianity is the most anthropocentric religion the world has seen."[27]

There is, however, a countercurrent of thought powerfully and discernibly running in the text of Genesis itself, however little representation it may have enjoyed in subsequent theology and popular Christianity. Within the general outlines of the traditional Scriptural world view, nonhuman species may have intrinsic value because they are parts of God's creation and God has conferred intrinsic value upon them, either by creating them or by a secondary fiat.[28]

The God of the Judeo-Christian tradition is transcendent, not immanent. The hypothesis of such a God therefore permits us to conceive of intrinsic value as determined objectively, i.e., from some point of reference outside human consciousness. From God's point of view, we may imagine, the creation as a whole and all its parts are "good." Everything may not seem good from a subjective human perspective—poison ivy, mosquitoes, rattlesnakes—but they are all "God's creatures" and therefore good in His "eyes."

It was upon just this theological-metaphysical ground that John Muir argued for "species rights." Notice how closely Muir's assertion of a natural right to existence for a vermin species is followed by appeal to God as a more objective axiological reference point: "Again and again, in season and out of season, the question comes up, 'what are rattlesnakes good for?' as if nothing that does not rightly make for the benefit of man had any *right to exist*; as if our ways were *God's ways*."[29] Muir repeatedly presses this theocentric orientation. All creatures, he urges, "are part of God's family, unfallen, undepraved, and cared for with the same species of tenderness and love as is bestowed on angels in heaven or saints on earth."[30]

How can we square Muir's and Ehrenfeld's interpretation of the axiology of Scripture with White's? A close reading of Genesis, in fact, discloses two different and even contradictory messages respecting the appropriate place and role of people in relation to the rest of creation. This should not be surprising in view of the modern discovery that Genesis as we have received it is woven together from three main narrative strands (designated as J, E, and P) all of different provenance.[31] J, the Yahwist strand, is by scholarly consensus the oldest, dating from the ninth century B.C., and P, the Priestly narrative, the most recent, composed in the fifth century B.C.[32]

The Priestly version of Creation (Genesis 1:2-4), in comparison with the Yahwist version, presents an orderly, rational "quasiscientific" account of the "evolution" of the cosmos. When reduced to its abstract moments—a primal unity (void, waters), separation of opposites (light/darkness, above/beneath), and serial production of living beings (plants, animals, people)—it is in form identical with the general outline of creation in fifth-century Ionian Greek natural philosophy. And like its contemporary Greek counterpart, the natural philosophy of the P strand of Genesis exhibits a distinct tendency toward humanism: man is created in the image of God and given dominion over the rest of creation and charged to subdue it.[33]

In the Yahwist creation myth (Genesis 2:4-4:26), the less "scientific" sequence of creation goes: man, then plants, then animals—and man's role is decidedly different. Adam is charged not to subdue and have dominion over the creation, but to "dress the garden and keep it" (Genesis 2:15). If he is not, in Aldo Leopold's terms, "a plain member and citizen" of Eden, neither is he its conqueror or master. Rather, Adam's role is to be custodian or steward of the creation.

Genesis-J thus seems quite clearly to imply that God cares for the creation as a whole (as "one great unit" in Muir's words) and for its several parts equally. The mastery of *Homo sapiens* over other species, the J narrative appears to assert, is a sign of the fallen and cursed condition of *Homo sapiens*, not of a privilege ordained by God. The assumption by humans of a self-centered or homocentric value orientation, indeed, seems radically to have unsettled the balance and order of the creation. Some animals and plants were enslaved (i.e., domesticated), those for which people could find no use were declared worthless, and those which confounded human purposes or made human life less comfortable or secure were declared pests and vermin and were put on an agenda for extermination.

There follows immediately the story of the destructive flood and of Noah, the original species conservator, for whom Ehrenfeld names his principle of "species rights."

Intrinsic value for nonhuman species based upon Ehrenfeld's Noah Principle and the metaphysic of Genesis-J would devolve upon species per se, not specimens. Individual beings come and go, each after its kind, while the created forms, species, persist. The destruction of species, though not of individuals, therefore, would be a denial of divine fiat by man. After all, Noah, following God's orders, did not attempt to save every individual living thing he could; rather he took specimens aboard the ark in male-female pairs so as to preserve their species. From this stock the creation could be restored, complete and intact. As the modern descendants of Noah, we ought, presumably, also to be more concerned with preserving species, with the value of species per se, and less with specimens and with individual nonhuman rights.

IV. HOLISTIC RATIONALISM

In the theistic moral metaphysic outlined in the previous section, God is the sole legitimate arbiter of value. It is not clear, however, how God goes about determining the value of nonhuman natural entities. On the one hand, it seems absurd to suppose that God has self-interests of any kind and, like man, determines value in relation to them, or that He is in need or want of anything or could be benefited or harmed in any way. God, we must suppose, is not injured so much as defied when His creation is altered or parts of it (species) are destroyed. It seems equally absurd, on the other hand, to suppose that God is arbitrary, that He simply and whimsically values the smallpox virus, the tsetse fly, and all the other forms of life that people, for the most part, find life-threatening, annoying, loathsome, or inconvenient. God must have followed some axiological principle(s) in deciding what to create and thus to confer value upon.

This line of thought, pursued far enough, separates value, classically called "the Good," from God. God Himself, from this perspective, is no longer the primary axiological reference point, since God now is thought to be determined or at least persuaded by some impersonal axiological principle, the Good.

The Good was classically conceived to be something "objective," i.e., independent of both divine and human interests, preferences, or desires.

The intrinsic value of nonhuman species could thus conceivably be grounded in an objective, impersonal principle of value as the primary axiological reference point.

One philosophical tradition, perhaps going all the way back to Plato, locates value or goodness in certain formal characteristics of systems or organized wholes. On the nature of the Good Plato was more suggestive than explicit. Recent Platonic scholarship inclines to the view that by "the Good" Plato meant a formal principle of order of the highest degree of generality, and by "order" meant formal logico-mathematical design.[34] A good house or ship is one that is well ordered, i.e., its parts are measured, proportioned, and fitted together according to a rational design; the goodness of body (health), of soul (virtue), of society (justice), and of the cosmos as a whole (literally, the world-order) is similarly defined.[35]

In the early modern period Leibniz more clearly or at least more explicitly defined what he took to be the objective, impersonal principle of value. Musing on why God chose just this world to create, Leibniz concludes that this one must be the best of all possible worlds. The enormous quantity of vice, pestilence, and calamity in the actual world makes this an outrageous statement from a homocentric point of view. It provoked Voltaire, indeed, to write *Candide* to illustrate the opposite thesis. But Voltaire's satire was beside the point, since by "best" Leibniz meant logico-mathematical elegance, not the absence of human frustration and suffering. According to Leibniz the Good that God had in view when choosing among possible worlds was "the greatest possible variety, together with the greatest order that may be; that is to say, . . . the greatest possible perfection."[36] Leibniz says that "God, however, has chosen the most perfect [world], that is to say the one which is at the same time the simplest in hypotheses and the richest in phenomena."[37]

Similarly, in contemporary conservation literature one sometimes finds biological diversity and/or complexity posited as a good in itself.[38] The most well-known application to ecological conservation of the general theory that the formal properties of natural systems—order, parsimony, harmony, complexity, and variety—are objective intrinsic values is the summary maxim of Aldo Leopold's "land ethic": "A thing is right when it tends to preserve the integrity, stability and beauty of the biotic community. It is wrong when it tends otherwise."[39]

Leopold makes no deliberate effort specifically to explain or defend his cardinal moral precept. A philosophical development of his ideas has recently been attempted by Peter Miller. Miller quite correctly points out that "most modern theories of . . . value, and indeed many classical ones, are psychologically [i.e., subjectively] based. They differ from one another just in the psychological phenomena they select as values."[40] Miller attempts to go beyond the orbit of all such subjective, homocentric theories by positing "richness" as an irreducible, objective, intrinsic value. While Miller very fully characterizes or describes "richness" ("the richness of natural systems [consists of] their inner and outer profusion, unity," etc.), he does not adequately explain why richness should be valued *for its own sake*, or, more concretely, why a diverse, complex, and stable biota is *intrinsically* better than a simple, impoverished, or unstable one. The value of "richness" is certainly explicable instrumentally: a biologically rich world is more satisfying and more secure than an impoverished world, but these are clearly homocentric concerns.

The nonhomocentric explanation of the goodness of order and variety which eludes Miller depends, in Leibniz's account as mentioned above, upon the hypothesis of God, certain assumptions about God's psychology, and the principle of sufficient reason.[41] God, as Leibniz forthrightly declares, has the tastes of a classical or early modern mathematician and natural philosopher.[42] Being infinitely rational, God prefers a logico-mathematically elegant world to one that is inelegantly designed. However, in the course of this Leibnizian explanation of the goodness of "richness," the alleged objective value of richness is reduced to a subjective preference—God's preference, to be sure, not ours, and a constitutional preference, not an arbitrary one, but a subjective preference nonetheless.

In any case, a persistent strain of Western axiological thought with the best philosophical pedigree posits an objective, impersonal Good and, further, characterizes or describes the Good in terms of formal elegance or logico-mathematical perfection: maximum economy of premises, axioms, or fundamental laws; maximum variety or diversity of implications or resulting phenomena; and consistency, order, or "harmony." Historically the universe has been found to exemplify these characteristics and thus to be objectively and impersonally good, even though from a subjective, homocentric standpoint it contains many "evils" which cause much human suffering. More recently the "biosphere" or global "ecosystem" has been found to exemplify similar characteristics and thus felt to be "good in itself," even though it may not be altogether accommodating to human interests.

Earth's biosphere is indeed an elegant system. The basic biological "laws" from which all its diversity and complexity result are wonderfully parsimonious. And while the relationships among species are many, intricate, and sometimes quite amazing, it seems they are comprehensible in terms of a relatively few basic chemical, physical, and perhaps topological processes.

From the rationalistic perspective, the system itself (the biosphere as a whole and/or its several biomes and integrated ecosystems) is valuable per se or at least exemplifies or embodies the Good. Therefore, from this perspective, species taken separately are not intrinsically valuable. However, since the intrinsically valuable biosphere is not some mystical or transcendental whole, but a systemic whole (i.e., a whole by virtue of the functional integration of its parts), its integrity, complexity, stability, variety, in a word, its inherently good or intrinsically valuable richness depends, obviously, on the continued existence of its component parts, i.e., its full complement of species. From the rationalist perspective, therefore, the preservation of species, as the *sine qua non* of the preservation of biotic diversity, ought to be pursued quite apart from the instrumental value preservation may have in relation to human interests.

Or at least so it would seem. However, if one defends one's intuition that biological impoverishment is objectively wrong by positing organic richness as objectively good, one might well be accused of temporal parochialism and a very subtle form of human arrogance. Considering our time as but an infinitesimal moment in the three and one-half billion year tenure of life on planet earth (let alone the possibility that earth may be but one of many planets to possess a biota), man's tendency to destroy other species might be viewed quite disinterestedly as a transitional stage in the earth's evolutionary odyssey. The Age of Reptiles came to a close in due course (for whatever reason) to be followed by the Age of Mammals. A holistic rationalist could not regret the massive die-off of the late Cretaceous because it made possible our yet richer mammal-populated world. The Age of Mammals may likewise end. But the "laws" of organic evolution and of ecology (if any there be) will remain operative. Nonhuman life would go on even after nuclear holocaust. In time speciation would occur and species would radiate anew. Future "intelligent" forms of life may even feel grateful, if not to us then to their God (or the Good), for making their world possible. The new Age (of Insects, perhaps) would eventually be just as diverse, orderly, harmonious, and stable and thus no less good than our current ecosystem with its present complement of species.

With friends like the holistic rationalists, species preservation needs no enemies.

V. Conativism

In sharp contrast to the objective, holistic orientation of moral thought in the ancient and early modern period, Western moral thought since the Enlightenment has been singularly narcissistic. Kenneth Goodpaster has argued that the "impotence" of modern Western ethics in the face of contemporary environmental problems is due to the fact that the two main modern schools of moral philosophy, deontological and utilitarian, assume egoism as an unquestioned given and then generalize to a larger set of intrinsically valuable "others."[43] The process of generalization begins by identifying an essential *psychological* characteristic that makes oneself, in one's own eyes, intrinsically valuable. According to Kant, founder of the deontological school, the characteristic is reason or rationality, and according to Bentham, founder of the utilitarian school, it is sentiency or the capacity to experience pleasure and pain. Egoism is then transcended by discovering the same characteristic in a select class of beings outside oneself.[44] It is revealing that both Kant and Mill, Bentham's protégé, invoke the Christian Golden Rule—love thy neighbor *as thyself*—as the perfect summary statement of their moral philosophies.

Kant's moral metaphysic limits intrinsic value to rational beings. Therefore, Kant's moral metaphysic is unsupportive of intrinsic value for nonhuman beings, either individually or collectively as species. Kant, indeed, directly stated that nonhuman living beings were of instrumental value only.[45]

Bentham's axiology is more inclusive than Kant's and is, in fact, the metaphysical foundation of the contemporary animal liberation/animal rights movement. Bentham himself recognized that the pleasure and pain of sentient animals must be taken into consideration no less than that of human beings,[46] but until very recently mainstream utilitarianism limited moral consideration to human welfare only. The contemporary animal liberation/animal rights philosophy gains much of its persuasive force from simply insisting that utilitarianism be put into practice in a logically consistent and intellectually honest way.

The core moral metaphysic of utilitarianism and of the animal liberation/animal rights movement is inadequate, however, to address massive species extinction.[47] In fact, it could under a certain extreme interpretation make matters worse. First, animal liberation excludes plants from moral consideration, shifting the burden of support for the rights-holding human-animal community onto plant species. Second, animal liberation/animal rights provides no philosophical basis for concern for species qua species, as Peter Singer, a leading animal liberation theorist, openly admits.[48] Animal liberation is concerned with the psychological well-being of individual animals, domestic no less than wild; its aim is to reduce individual animal suffering. A species qua species cannot experience pleasure or pain and thus upon Benthamic principles is entitled to no moral consideration. Since wild animals often suffer considerably in their natural habitats from extremes of cold, drought, starvation, disease, and predation, the animal liberationists' program of reducing individual animal suffering might achieve a "final solution" by the deliberate, painless extinction of all sentient nonhuman animal species.[49] Or perhaps, as Mark Sagoff once remarked, from the point of view of animal liberation, the best thing for wild animals would be relocation in zoos, where they could be cared for and protected from the suffering inflicted on them by the elements and by one another.[50]

While Bentham's utilitarian moral metaphysic is more inclusive than Kant's, it proves to be useless as a foundation for the intrinsic value of nonhuman species. However, certain historical modifications of Kant's deontological ethic, surprisingly, may be of some service in building a nonhomocentric case for species preservation.

The neo-Kantian voluntarist ethical tradition which begins with Schopenhauer substitutes conation (the "will-to-live") for reason as the essence of the self.[51] Conation or the "will-to-live," of course, is far more universal than reason, and at the very least resides in every living thing. (Schopenhauer thought that it also was the "kernel" of everything right down to elemental matter—of which inertia and gravity were the striving—but more recent theorists in the conativist or voluntarist tradition are not so generous.) Generalizing from conation as the essence of self, it follows that all beings which are "manifestations" of the "will-to-live," i.e., at the very least all living things, have intrinsic value.

Among recent exponents of conativism, Albert Schweitzer's "reverence-for-life" ethic exhibits the clearest traces of Schopenhauer's influence together with an explicit illustration of the modern method of generalizing from egoism to altruism.[52] Those Anglo-American moral philosophers who base moral standing (and sometimes rights) upon "interests" construed in the broadest sense are also, though they are usually not so identified, voluntarist fellow travelers. The term "interest" is, of course, ambiguous. Setting aside nonpsychological, nondispositional senses (e.g., the financial sense), "interest" has been construed in three principal ways. One may have an interest in the sense of having one's attention engaged. This may be called the cognitive sense of "interest." If having interests means having one's attention engaged and the capacity for having interests is the criterion of intrinsic value, then only those human beings and the higher vertebrates with cognitive capacities are intrinsically valuable.[53] The capacity for having interests has been somewhat more broadly construed by animal liberationists in terms of "sentiency," the capacity for experiencing pleasure and pain.[54] This may be called the *hedonic* sense of "interest." Joel Feinberg has construed interests in a sense broader still to mean a *conative* capacity:

A mere thing, however valuable to others, has no good of its own. The explanation of that fact, I suspect, is that mere things have no conative life; neither conscious wishes, desires and hopes; nor urges and impulses; nor unconscious drives, aims, goals; nor latent tendencies, directions of growth, and natural fulfillments. Interests must be compounded somehow out of conations.[55]

Though Feinberg himself does not appear to appreciate this consequence, under this view of "interests" plants as well as animals may have interests and thus intrinsic value, because, though plants may not have "conscious wishes, desires, and hopes," they have "tendencies, directions of growth, and natural fulfillments." Goodpaster explicitly draws the implication, from Feinberg's discussion of interests, that plants, too, are in Goodpaster's terminology "morally considerable."[56] On this basis, Goodpaster defends a "life principle" of moral considerability which includes all living things.

The Schopenhauer-Schweitzer reverence-for-life ethic and the Feinberg-Goodpaster life principle ethic avoid some of the untoward characteristics of the Benthamic animal liberation moral metaphysic as applied to the question of species extinction. Clearly plants are included within the moral community as well as animals. And since the essential capacity identified as the criterion of moral considerability is conativity, not sentiency—a thrusting, striving, driving, developmental tendency or direction (whether conscious or unconscious)—the life principle and reverence-for-life ethics do not have the effete and prophylactic connotations of the animal liberation/animal rights hedonic ethic. They suggest, to me at any rate, that living things should be left alone to fulfill their natural urges, drives, and developmental and reproductive sequences, or to struggle, fight, and die in the attempt rather than be coddled, sheltered, protected, anesthetized, or otherwise "saved." One's cardinal duty is not to interfere, to live and let live.

Species per se, however, are no more conative than they are sentient or rational, hence species per se are not intrinsically valuable from this point of view. Species qua species may however be the incidental beneficiaries of an ethic directed toward the preservation of individual living beings since the reverence-for-life and life principle ethics would surely imply a far less callous, mindlessly destructive approach to the biota than that which currently prevails.

Indeed, one of the principal problems with the conation-based moral metaphysic is that, if rigorously practiced, it would seem to require a restraint so severe that it would lead if not to suicide by starvation, at best to a life intolerably fettered. Schopenhauer, always intellectually honest, was prepared to accept these practical consequences. His more recent exponents recognize them as practical consequences, but treat them as a problem to be somehow got round. As Schweitzer remarks, "It remains a painful enigma how I am to live by the rule of reverence for life in a world ruled by creative will which is at the same time destructive will"; and Goodpaster remarks, "the clearest and most decisive refutation of the principle of respect for life is that one cannot *live* according to it, . . . we must eat, experiment to gain knowledge, protect ourselves from predation (macroscopic and microscopic), . . . to take seriously the criterion of considerability being defended, all these things must be seen as somehow morally wrong."[57]

Goodpaster addresses this problem by means of a formal distinction which is, despite his claim to the contrary, largely vacuous. Since we are subject to certain "thresholds of moral sensitivity," ideally we may acknowledge the "rights" of all living things to exist, but practically we may be unable to live on such terms. Such ideals are "regulative," not "operative."[58] We are thus left paying lip service to an impractical ideal while day-to-day life goes on pretty much business as usual.

Schweitzer hints at a decision procedure which might put some teeth in the reverence-for-life ethical ideal: "Whenever I injure life of any kind, I must be quite clear as to whether this is necessary or not. I ought never to pass the limits of the unavoidable, even in apparently insignificant cases."[59] But the rule "never to pass the limits of the unavoidable" is very vague and indeterminate. The destruction of critical habitat for an endangered species may be judged "unavoidable" by a consortium intent on developing its "resources" at a reasonable profit. The extinction of several species of great whales may be "unavoidable" if the whaling industry is to recover its capital investment. More explicit criteria are needed if a moral theory according intrinsic value to all living things is to be at once "operative" (in Goodpaster's sense) and practical or livable.[60]

If our society were to acknowledge and institutionalize a reverence-for-life or life principle ethic to the same extent that it has acknowledged and institutionalized an ethic based upon justice and human equality, things would be as different from what they are today for nonhuman forms of life as contemporary human life (in most democratic societies at least) differs from the oppressive conditions of the imperial and feudal past. Still, there is no logical link as far as I can see between a concern for the intrinsic value of *individual* plants and animals and a concern for *species* preservation. To be sure, a species survives only if its representative specimens are allowed to survive and to reproduce successfully. However, according to the conative theory of intrinsic value, individual living things are in principle of equal value, while species preservationists set a much higher value, for example, on an individual furbish lousewort (a "mere" plant, but a precious custodian of unique genetic material) than on an individual whitetail deer, a commonplace mammal. A life-respecting society might significantly slow the rate of species extinction, but species preservation would be, nevertheless, an *incidental* consequence, a side effect. This is, however, the best that the prevailing structure of modern moral philosophy can do.[61]

VI. Bio-Empathy

There remains a modern moral metaphysic which has been largely ignored or dismissed by the philosophical community, but which has survived largely in biological discussions of moral or moral-like phenomena. Hume's grounding of morality in feeling or emotion has been the basis for several recent attempts to explain the intrinsic value of other species. According to Hume, one may have a strong emotional attachment to one's own interests, but such an attachment is entirely contingent. It is possible, indeed, that one may also have strong feelings for the interests of other beings.[62] Sometimes these overcome the self-regarding passions and issue in behavior which we praise as "heroic," "noble," or "saintly" (or condemn as "foolhardy" or "daft").

Hume's famous sharp distinction between fact and value, his is/ought dichotomy, has made his moral metaphysic more appealing and useful to scientists interested in moral phenomena than any other philosophical analysis of ethics, since in science nature is conceived to be an objective and, more to the point so far as our interests are concerned, value-free system. From the scientific point of view, nature throughout, from atoms to galaxies, is an orderly, objective, axiologically neutral domain. Value is, as it were, projected onto natural objects or events by the subjective feelings of observers. If all consciousness were annihilated at a stroke, there would be no good and evil, no beauty and ugliness, no right and wrong; only impassive phenomena would remain. Accordingly, it has been characteristic of evolutionary biological thought about moral phenomena to follow Hume (whether deliberately or not) and treat moral valuation and behavior as both subjective and affective.

One of the more conspicuous problems for an evolutionary biological account of animal behavior is this: How is it possible to account for the existence of something like morality or ethics among human beings and their prehuman ancestors in a manner consistent with evolutionary theory? One would suppose, given the struggle for existence, that hostile, aggressive traits would be of great advantage to individuals in competition with one another for limited resources and that therefore such traits would be represented in ever increasing magnitude in future generations. As time goes on we should see less inclination toward "moral" behavior, rather than, as the history of civilization seems to indicate (though cynics might well contest this point), more. At this late date, in any case, all human beings, indeed all animals, should be thoroughly rapacious and utterly merciless. Kindness, pity, generosity, benevolence, justice, and similar dispositions should have been nipped in the bud as soon as they appeared, winnowed by the remorseless and impersonal principle of natural selection.

Charles Darwin himself tackled this problem in *The Descent of Man*.[63] He begins with the observation that for many species, and especially mammals, prolonged parental care is necessary to ensure reproductive success. Such care is motivated by a certain strong emotion which adult mammals (in some species perhaps only the females) experience toward their offspring—parental love. Selection for this capacity would affect a species' psychological profile since it would strongly contribute to inclusive fitness (not necessarily prolonged individual survival, so much as reproductive success).

Once established, Darwin argued, the "parental and filial affections" permitted the formation of small social units originally consisting, presumably, of parent(s) and offspring. The survival advantages to the individual of membership in a protective social unit, like a family group, are obvious and would tend to conserve slight variations of the parent-child emotional bond, such as affection for other kin—siblings, uncles, aunts, cousins, and so on. Those individuals in whom these affections were strongest would form the most closely knit family and clan bonds. Now, these and similar "social sentiments" or "social instincts," such as "the all-important emotion of sympathy," Darwin reasoned, "will have been increased through natural selection; for those communities which included the greatest number of the most sympathetic members would flourish best, and rear the greatest number of offspring."[64]

As family group competes with family group, ironically, the same principles which at first would seem to lead to greater intolerance and rapacity lead instead to increased affection, kindness, and sympathy, for now the struggle for limited resources is understood to be pursued collectively, and groups with "the greatest number of the most sympathetic members" may be supposed to out-compete those whose members are quarrelsome and disagreeable. "No tribe," Darwin tells us, "could hold together if murder, robbery, treachery, etc., were common; consequently, such crimes within the limits of the same tribe 'are branded with everlasting infamy'; but excite no such sentiment beyond these limits."[65] Indeed, beyond these limits, it remains biologically important for the passions of aggression, rage, and bloodlust to come into play.

Not only was there selective pressure for *more intense* sympathy and affection within group boundaries, there was selective pressure for more widely cast social sentiments, since in competition among the most internally peaceable and cooperative groups the larger will win out. "As man advances in civilization, and small tribes are united into larger communities, the simplest reason would tell each individual that he ought to extend his social instincts and sympathies to all the members of the same nation though personally unknown to him [and unrelated to him genetically]."[66]

Unlike both the (Benthamic) utilitarian and (Kantian) deontological schools of modern moral philosophy, the Humean-Darwinian natural history of morals does not regard egoism as the only genuine and self-explanatory value. Selfishness and altruism are equally primitive and both are explained by natural selection. Self-assertion and aggressiveness are necessary for survival to reproductive age and to reproductive success, but so are caring, cooperativeness, and love.

Darwin's account of the origin and evolution of morals obviously involves the current biological anathema of "group selection," i.e., natural selection operating with respect to groups rather than to individual phenotypes who are the immediate carriers of these genes.[67] A more rigorous theoretical account of social-moral phenomena has recently been provided by social evolutionary theorists.[68] Darwin's classical account, however, is an indispensable ingredient in the theoretical structure of Aldo Leopold's "land ethic" (which contains a plea for the "biotic right" of other species to exist) and it is the basis for Paul and Anne Ehrlich's argument for the "rights" of species to exist, as well.

Leopold's biological description of an ethic as "a limitation on freedom of action in the struggle for existence"[69] at once locates ethics in a Darwinian context and suggests the evolutionary paradox presented by ethical phenomena. His resolution of the paradox is Darwin's in a nutshell. An ethic, according to Leopold, "has its origin in the tendency of interdependent individuals or groups to evolve modes of cooperation."[70] Leopold, following Darwin, believes that growth in the extent and complexity of human ethics, what he calls the "ethical sequence," has paralleled and facilitated growth in the extent and complexity of human societies. Leopold envisions the land ethic as the next "step" in this pattern of social-ethical expansion. Social evolution has recently achieved a worldwide human society and ethically we have achieved, corresponding to this social condition, the ideal of universal "human rights." Ecology, Leopold points out, represents the relationship of human and nonhuman organisms in the natural environment by means of a "community concept." Were this ecological idea of a "biotic community" to become widely current, Leopold foresees the emergence, correlatively, of a "land ethic" or "ecological conscience."

Many biologists have come to see the world through a prism of evolutionary and ecological theory. Moreover, as scientists they participate in a more general "Copernican" world view. The earth is perceived as a very small, lush, blue-green island in a vast desert sea of space. Biotas may exist on other planets, but these would be genuinely "foreign," "alien," in comparison to which earth's organisms are all literally kin. If Darwin is correct that the perception of another being as a family and/or community member triggers in us certain instinctive emotional responses, and if all the denizens of the "small planet" earth are so perceived, then something like Leopold's land ethic may become an operative ideal for future civilization.

These conceptual elements are all present in abbreviated form in Paul and Anne Ehrlich's impassioned appeal for "species rights": "Our fellow passengers on Spaceship Earth, who are quite possibly our only living companions in the entire universe, *have a right to exist.*"[71] The phrases "Spaceship Earth" and the "entire universe" evoke the Copernican perspective and "fellow passengers" and "living companions" evoke the evolutionary-ecological world view.

The Ehrlichs go on to provide a more extended rationale for "species rights," and though they do not mention Darwin by name, their understanding of the origin and evolution of morals is step for step Darwin's own and their projection of future moral evolution to include other species recapitulates the next step in the "ethical sequence" of Leopold's land ethic:

> Along with other ecologists, we feel that the extension of the notion of "rights" to other creatures . . . is a natural and necessary extension of the cultural evolution of *Homo sapiens*. . . . From an original concern only with the family or immediate group there has been a steady trend toward enlarging the circle toward which ethical behavior is expected. First the entire tribe was included, then the city-state, and more recently the nation. In this century concern has been extended in many groups to encompass all of humanity. . . . In the last hundred years, the ranks of those in the United States and Europe advocating compassion for, and unity with, the rest of the natural world have swollen considerably.[72]

Moral metaphysics "from the side of natural history," as represented by Hume, Darwin, Leopold, and the Ehrlichs, differs at several key points from moral metaphysics from the side of philosophy. I have already pointed out that in the biological tradition egoism is not the only irreducible, primitive value. Affection and sympathy, the "moral sentiments," are on equal footing with "self-love." Further, there is no preoccupation with psychological states as intrinsically valuable in and of themselves—no special concern with pleasure and pain, reason and knowledge, interests, or a hierarchy of beings determined by psychological complexity. The importance of this difference and the next cannot be stressed enough. The value of organisms is not gauged by how they feel, nor by how they make humans feel, although their value ultimately depends upon certain "intentional" mammalian affections. And, while the two mainstream modern philosophical accounts and the natural history account of morality can provide for the intrinsic value of individual nonhuman organisms, the philosophical accounts grant moral standing for individuals only, while the natural history account makes possible moral status for wholes. Hume, for example, recognizes a distinct sentiment which naturally resides in human beings for the "publick interest."[73] Darwin recognizes affection not only for "fellows" but for "family" and "tribe," i.e., in general, "the good or welfare of the community."[74] Leopold says that his land ethic would require of *Homo sapiens* "respect for the [biotic] community as such."[75] The Ehrlichs also talk about species qua species, as well.

Thus I think we have found, at last, an axiology which faithfully articulates and adequately grounds the moral intuition that nonhuman species have "intrinsic value." They may not be valuable *in* themselves, but they may certainly be valued *for* themselves. According to this expanded Humean account, value is, to be sure, humanly conferred, but not necessarily homocentric. We certainly experience strong self-oriented feelings and appraise other things in reference to our human interests. But we experience certain distinct disinterested affections as well. We can foster, for example, the welfare of our own kin at considerable cost or even sacrifice to ourselves. We are capable of a disinterested sympathy and selfless charity to persons unrelated and unknown to us. According to Hume, the "intrinsic value" we attribute to all human beings is a projection or objectification of this "sentiment of humanity."

The philosophical and popular disagreement about which beings are intrinsically valuable, though all value is itself affective, is, according to this theory, a matter of *cognitive* rather than affective differences. The human capacity for the moral sentiments upon which intrinsic value depends is fairly uniform (because it is a genetically fixed psychological characteristic like sexual appetite) and roughly equally distributed throughout the human population. To whom or to what these affections are directed, however, is an open matter, a matter of cognitive representation—of "nurture," not "nature." A person whose social and intellectual horizons are more or less narrow regards only a more or less limited set of persons and a more or less local social whole to be intrinsically valuable. To perceive nonhuman species as intrinsically valuable involves, thus, not only the moral sentiments, but an expansive cognitive representation of nature.

The Humean/Darwinian bio-empathetic moral metaphysic, based upon naturally selected "moral sentiments," provides a theory according to which species qua species may have "intrinsic value." That is, they may be valued for themselves. Because the theory is humanly grounded, though not humanly centered, it does not impel us toward some detached and impersonal axiological reference point and thus submerge the value of the present ecosystem in a temporally and spatially infinite cosmos, as Holistic Rationalism does. Our social affections are extended to our fellow members and to the social whole of which we are part. The tribesmen who stand helplessly by and witness the "extinction" of their culture, as so many nineteenth-century Native Americans unfortunately had to do, take little comfort in knowing that another cultural order will replace their own. Similarly, this is the biotic community of which we are a part, these are our companions in the odyssey of evolution, and it is to them, not to any future complement, that our loyalties properly extend.

Hume's grounding of morality in feeling or emotion has usually been regarded by the philosophical community as leading inevitably to an irresponsible ethical relativism.[76] If good and evil, right and wrong, are, like beauty and ugliness, in the eye of the beholder, then there can be no moral truths. We could no more reject as mistaken the opinion that matricide, say, is good, than the opinion that Picasso's Cubist paintings are ugly.

While Hume's theory of morality is certainly an emotive theory, it does not necessarily collapse into emotive relativism. Hume provides for a functional equivalent of objective moral truths by what may be called a "consensus of feeling." The human psychological profile in certain crucial respects is standardized, fixed. Unlike aesthetic judgments, which notoriously vary widely from culture to culture and within the same culture from person to person, moral judgments (allowing for certain peripheral divergencies) are both culturally and individually invariant. Christian cultures may regard polygamy with horror while Muslim cultures may approve it. Still, all cultures abominate murder, theft, treachery, dishonesty, and the other cardinal vices. Certainly individuals differ in the degree to which they are endowed with the moral sentiments. Still, just as we can speak of certain normal physical proportions and conditions among human beings, while allowing for all sorts of variations, so we can speak of a certain normal human affective profile, while allowing for all sorts of variations. Some people are tall, others are short, and both the tall and the short are normal. Then there are giants and midgets. Similarly, some people are overflowing with moral sentiments while others experience them far less intensely and are more possessed by self-love. Depraved criminals, for example, exceed the limits of normality. They are the psychological equivalent of the physically freakish. Their emotional responses are not untrue, but, by the human consensus of feeling, they are "wrong," morally, if not epistemically.

For Hume the "universality" of human moral dispositions was an ad hoc fact. Darwin completed Hume's theory by explaining how such a standardization came about. Like the complex of normal human physical characteristics, normal human psychological characteristics, including the moral sentiments, were fixed by natural (and perhaps by sexual) selection.

Still it may seem defeating to say that the nonutilitarian value of other forms of life is ultimately emotional, that it rests upon feeling, that species are valuable and we ought to save them simply because we have an affection for them. This would be defeating if there were some viable alternative and if emotivism implied moral relativism. But according to the Humean-Darwinian axiology, the only tenable axiology from the general perspective of traditional normal science, all value is affective. The intrinsic value we attribute to individual human beings and to humanity expresses only our feelings for co-members of our global village and for our human community. I remain convinced therefore that the Humean-Darwinian moral metaphysic is, intellectually, the most coherent and defensible axiology and, practically, the most convincing basis for an environmental ethic which includes intrinsic value for nonhuman species.

VII. Conclusion

In the foregoing discussion I have stressed the importance of the question, "Why try to preserve threatened and endangered species?" There are good "utilitarian" or "homocentric" reasons for preserving all or almost all existing species. Other species contribute to human well-being as performers of vital services, as resources, and as functional components in the global (human) life-support system, "Spaceship Earth." Frequently one also finds a distinctly "nonutilitarian" or "nonhomocentric" argument for species preservation, viz., that we have a moral obligation not to extirpate species or, more commonly, that other species, no less than we, have a right to exist, a right to a share of life on the planet. The nonutilitarian or nonhomocentric ethical argument for species preservation is said (by some at least) to be the most compelling reason for species preservation, but, paradoxically, it has been the least well articulated. Accordingly, my primary goal has been to explore and evaluate possible conceptual bases of the nonutilitarian or nonhomocentric argument for species preservation, and more particularly, to analyze and evaluate the assertion of rights on behalf of species.

The concept of "species rights" is not without its problems. Because of its conceptual difficulties, from a philosophical point of view, it would be better abandoned altogether. But philosophers have little influence on the vagaries of popular usage. The assertion of "species rights" upon analysis appears to be the modern way to express what philosophers call "intrinsic value" on behalf of nonhuman species. Thus, the question, "Do nonhuman species have a right to exist?" transposes to the question, "Do nonhuman species have intrinsic value?" There are several distinct moral metaphysics which might yield a positive answer to this question: J-Theism, Rational Holism, Conativism, and Bio-empathy.

Of these distinct types of moral theory, J-Theism and Bio-empathy appear to me to provide most effectively for the intrinsic value of other species. Each has wide appeal to different and complementary segments of the public, and each is relatively simple and straightforward.

Conativism most accords with prevailing biases in philosophical ethics, but because of its intractably "atomic" or "individualistic" ontology it can provide at best only incidentally for moral concern over vanishing species. Holistic Rationalism has some contemporary popular appeal and some contemporary philosophical representation, but it is more plausible in an ancient and early modern creationist context of thought. As a value theory it is so general, abstract, and impersonal that pressed to its logical extremes it might ill serve the cause of species preservation.

Only J-Theism unequivocally provides for objective intrinsic value for existing nonhuman species. The cognitive complex with which J-Theism is associated, the Judeo-Christian world view, is culturally well established and familiar. The greatest cultural competitor of the Judeo-Christian world view is Scientific Naturalism, with which the Bio-empathic axiology is conceptually and historically associated. Those unpersuaded by J-Theism, because of Judeo-Christianity's conflict with Scientific Naturalism, are likely, therefore, to be persuaded by Bio-empathy.

So, if the Western world's two main cultural belief systems, Judeo-Christianity and Scientific Naturalism, both provide for the intrinsic value of other species, why does the notion that nonhuman species have intrinsic value seem so foreign and why does it attract so much skepticism, opposition, and ridicule? Unfortunately for our nonhuman companions on the planet, the Judeo-Christian world view also harbors an axiology contradictory to J-Theism, namely P-Theism. P-Theism's moral metaphysic permits, if it does not require, the interpretation that human beings are morally privileged. In the P version of Genesis, human privilege is supported by the doctrine that God created human beings in His own image and favored them in His creation as rightfully holding "dominion" over nature. "Dominion" could be taken in several senses, one of which might imply a "steward" role for man in relation to nature, but it has more usually been taken to imply "mastery." The dogged insistence by many people that other forms of life have only instrumental value is probably traceable to this strain of thought in the Judeo-Christian tradition. Resistance to the notion of intrinsic value for other species in the scientific community, on the other hand, may be the result of residual acceptance of Judeo-Christian human chauvinism or may stem from the mistaken belief that since values, from a scientific point of view, are not wholly objective, they are therefore necessarily selfish or narcissistic, somehow unreal, or otherwise specious.

NOTES

1. See George M. Woodwell, "The Challenge of Endangered Species," in *Extinction Is Forever*, edited by Ghillian Prance and Thomas Elias (New York: New York Botanical Garden, 1977), p. 5.
2. See Thomas Eisner et al., "Conservation of Tropical Forests," *Science* 213 (1981): 1314, and Thomas E. Lovejoy, this volume.
3. See International Union for Conservation of Nature and Natural Resouces (IUCN), *Red Data Book* (Morges, Switzerland: IUCN, 1974) and Norman Myers, "An Expanded Approach to the Problem of Disappearing Species," *Science* 193 (1976): 198-201.
4. Norman Myers, *The Sinking Ark: A New Look at the Problem of Disappearing Species* (New York: Pergamon Press, 1979), p. 4. This seemingly preposterous rate is based upon the assumption that systematic deforestation of moist tropical forests could result in the loss of òne million species by the turn of the century (see Eisner et al., "Conservation of Tropical Forests"). Considering how close we are to 2000 A.D., the rate of 100 per day actually appears conservative. An average of more nearly 150 species extinctions per day would have to take place if one million species were to go extinct between now and the year 2000.
5. Eisner et al., "Conservation of Tropical Forests"; Myers, *Sinking Ark*, p. 5.
6. A. R. Wallace, *The Geographical Distribution of Animals* (London: Macmillan, 1876), p. 150.
7. This and the other questions in the paragraph are routinely posed rhetorical questions. Should the figure of 90 + percent be doubted, however, see David M. Raup, "Size of the Permo-Triassic Bottleneck and Its Evolutionary Implications," *Science* 206 (1979): 217-18.

8. See Normal D. Newell, "Crises in the History of Life," *Scientific American* 208 (1963): 76-92; David M. Raup and J. John Sepkoski, Jr., "Mass Extinctions in the Marine Fossil Record," *Science* 215 (1982): 1501-1503.

9. Two recent works, Myers, *Sinking Ark*, and Paul and Anne Ehrlich, *Extinction: The Causes and Consequences of the Disappearance of Species* (New York: Random House, 1981), are in large part convenient catalogues of utilitarian, or more accurately "homocentric," arguments for species preservation. Also see Bryan G. Norton, this volume, and Alastair S. Gunn, "Preserving Rare Species," in *New Introductory Essays in Environmental Ethics* (New York: Random House, 1984), pp. 289-335. Gunn provides a taxonomy and critical discussion of utilitarian or homocentric arguments for species preservation.

10. As a system of philosophical ethics, utilitarianism does not posit human happiness or human well-being as the *summum bonum*. Rather, Jeremy Bentham and John Stuart Mill, the founders of utilitarianism, declared that pleasure is good and pain is evil and that it is the duty of a moral agent to maximize the one and minimize the other no matter where located, i.e., no matter by whom experienced. Cf. Jeremy Bentham, *Introduction to the Principles of Morals and Legislation*, New Edition (Oxford: The Clarendon Press, 1823), chap. I, secs. I and X, and John Stuart Mill, *Utilitarianism* (New York: The Library of Liberal Arts, 1957), chap. 2. The implications of this view for animal liberation and the preservation of species are discussed below.

11. Woodwell, "The Challenge of Endangered Species," p. 5, and Howard S. Irwin, *Extinction Is Forever*, Preface, p. 2. Cf. also Michael Soule's comment in *Proceedings of the U.S. Strategy Conference on Biological Diversity, Nov. 16-18, 1981* (Washington, D.C.: Department of State Publication 9262, 1982), p. 61: "[I]t is regrettable that we must all pretend to be concerned exclusively with man and his welfare and put nearly all of our arguments for conservation for biological diversity in terms of benefit for man. [W]hen [will we] admit in public that conservation is not only for people, something most of us already admit in private [?]"

12. William Godfrey-Smith, "The Rights of Non-humans and Intrinsic Values," in *Environmental Philosophy*, edited by Don Mannison, Michael McRobbie, and Richard Routley (Canberra: Australian National University, 1980), p. 31, shares my suspicions: "Although environmentalists often use the rare herb argument, it seems to me that it is really only a lever; it does not express a very significant component of their thinking." See also Alastair S. Gunn, "Why Should We Care About Rare Species?" *Environmental Ethics* 2 (1980): 17-37.

13. Examples, in chronological order, are: John Muir, *Our National Parks* (Boston: Houghton Mifflin, 1901), p. 57, and *A Thousand Mile Walk to the Gulf* (Boston: Houghton Mifflin, 1916), p. 98; Aldo Leopold, *A Sand County Almanac* (Oxford: Oxford University Press, 1949), pp. 210, 211; Charles Elton, *The Ecology of Invasions by Animals and Plants* (London: Methuen, 1958), p. 144; David Ehrenfeld, "The Conservation of Non-Resources,"

American Scientist 64 (1976): 654; Bruce MacBryde, "Plant Conservation in the United States Fish and Wildlife Service," in *Extinction Is Forever*, p. 70; Ehrlich and Ehrlich, *Extinction*, p. 48; Roger E. McManus and Judith Hinds, eds., *The Endangered Species Act Reauthorization Bulletin* 1 (Washington, D.C.: Center for Environmental Education, Dec. 1981), p. 3.

14. Gunn, "Preserving Rare Species," p. 330.
15. Mark Sagoff, "On the Preservation of Species," *Columbia Journal of Law* 7 (1980): 64, claims that "we enjoy an object because it is valuable; we do not value it merely because we enjoy it.... Esthetic experience is a perception, as it were, of a certain kind of worth." For a similar judgment about aesthetic experience as applied to the question of species preservation see Lilly-Marlene Russow, "Why Do Species Matter?" *Environmental Ethics* 3 (1981): 101-12. William F. Baxter, *People or Penguins: The Case for Optimal Pollution* (New York: Columbia University Press, 1974), p. 5, however, turns this argument on its head: "Damage to penguins, or sugar pines, or geological marvels is, without more, simply irrelevant. One must go further ... , and say: Penguins are important because people enjoy seeing them walk about rocks. ..." Ehrenfeld, "The Conservation of Non-Resources," p. 654, discusses the aesthetic rationale for species preservation and concludes that "it is rooted in the homocentric, humanistic world view," since it appeals, finally, to what "is stimulating to man." He finds the aesthetic rationale incompatible with the "humility-inspiring discoveries of community ecology or with the sort of ecological world view, emphasizing the connectedness and immense complexity of man-nature relationships, that now characterize a large bloc of ecological thought." This is also the case, I think with Donald Regan's novel argument for the intrinsic value of experiences regarding nonhuman species (this volume). The "organic unity" of the "complex" consisting of a natural object, human knowledge of a natural object, and the human pleasure taken in that knowledge notwithstanding, upon Regan's argument nonhuman species remain only instrumentally valuable as *epistemic resources*; the value Regan finds in nonhuman species is formally the same as aesthetic value, since species are valuable according to his account as objects of epistemic experience rather than as objects of aesthetic experience. The putative intrinsic value he claims for nonhuman species is susceptible to reduction to mere instrumental value. As either aesthetic or epistemic objects, nonhuman species are valued only as means to an intrinsically valuable state of human consciousness or so either Baxter or Ehrenfeld might insist.
16. See, for example, Holmes Rolston III, "Are Values in Nature Subjective or Objective," *Environmental Ethics* 4 (1982): 125-51; and Don E. Marietta, Jr., "Knowledge and Obligation in Environmental Ethics: A Phenomenological Approach," *Environmental Ethics* 4 (1982): 153-62.
17. See for example, Christopher Stone, *Should Trees Have Standing? Toward Legal Rights for Natural Objects* (Los Altos: William Kaufman, 1974) for an "operational" definition. It may be worth noting that the Endangered

Species Act of 1973 confers rights upon specimens of endangered species according to Stone's operational criteria, although the Act does not specify "rights" per se and grounds its protection for endangered species exclusively in utilitarian terms.

18. See H.L.A. Hart, "The Ascription of Responsibility and Rights," in *Logic and Language*, edited by Anthony Flew (Garden City: Anchor Books, 1965), pp. 151-74.

19. John Rodman, "The Liberation of Nature," *Inquiry* 20 (1977): 108, agrees with this analysis of the popular preservationist usage of "rights": "To affirm that 'natural objects' have 'rights' is symbolically to affirm that all natural entities, including humans, have intrinsic worth simply by virtue of being." Nicholas Rescher, "Why Save Endangered Species?" in *Unpopular Essays on Technological Progress* (Pittsburgh: University of Pittsburgh Press, 1980), agrees that species per se cannot be coherently attributed rights. He also asserts that we have an ethical duty to save endangered species because they possess a metaphysical intrinsic value. He does not, however, undertake to provide a *theory* of intrinsic value or detail a metaphysics which conceptually grounds the intrinsic value of species.

20. David L. Hull, "A Matter of Individuality," *Philosophy of Science* 45 (1978): 335-60.

21. For a general discussion, see Michael Ruse, "Definitions of Species in Biology," *The British Journal for the Philosophy of Science* 20 (1969): 97-119. For a critical discussion of Hull's views see D. B. Kitts and D. J. Kitts, "Biological Species as Natural Kinds," *Philosophy of Science* 46 (1979): 613-22, and Arthur L. Caplan, "Back to Class: A Note on the Ontology of Species," *Philosophy of Science* 48 (1981): 130-40.

22. In the interests both of contemporary relevance and saving space I shall *not* discuss those classical moral metaphysics which might provide for the intrinsic value of nonhuman species, but which have few contemporary exponents. An example of one such theory is G. E. Moore's Intuitionism in which value or "goodness" is alleged to be an objective, but "non-natural," quality which one may discern by one's unaided moral sensibilities.

23. Ehrenfeld, "The Conservation of Non-Resources," p. 654.

24. Ibid., p. 655. Similar ideas are expressed in his "What Good Are Endangered Species Anyway?" *National Parks and Conservation Magazine* 52 (October 1978): 10-12; and *The Arrogance of Humanism* (New York: Oxford University Press, 1978), pp. 207-11.

25. See John Passmore, "The Treatment of Animals," *Journal of the History of Ideas* 36 (1975): 195-218, for a definitive discussion.

26. Lynn White, Jr., "The Historical Roots of Our Ecologic Crisis," *Science* 155 (1967): 1203-1207. White does not consider an alternative, environmentally more sympathetic interpretation of the verses in question, generally referred to as "stewardship." According to the stewardship interpretation of Scripture, man's superiority implies not only privilege but responsibility. For a scholarly elaboration and defense of a stewardship reading of Genesis 1:26-30 see

James Barr, "Man and Nature: The Ecological Controversy and the Old Testament," *Bulletin of the John Rylands Library* 55 (1972): 9-32.

27. Lynn White, Jr., "Historical Roots," p. 1205.

28. It should be kept in mind that the idea that human beings possess moral or natural rights was initially defended by John Locke in his quaint *First Treatise of Government* in Scriptural terms. God, according to Locke, conferred rights upon Adam and his descendants. In this connection we should remind ourselves of Thomas Jefferson's famous words in the Declaration of Independence: "all men . . . *are endowed by their Creator* with certain inalienable rights. . . ." Human worth and dignity thus were once commonly grounded in a theocentric moral metaphysic.

29. John Muir, *Our National Parks*, p. 57 (emphasis added).

30. Muir, *Thousand Mile Walk*, pp. 98-99.

31. Arthur Weiser, *The Old Testament: Its Formation and Development*, translated by D. Barton (New York: Association Press, 1961).

32. Ibid., p. 77.

33. See F. M. Cornford, *Principia Sapientia* (Cambridge: Cambridge University Press, 1952), chap. 11, for a detailed dicussion.

34. Cf. "The Tübingen School," most notably, H. J. Kramer, *Arete bei Platon und Aristoteles: zum Wesen und zur Geschichte der platonischen Ontologie* (Heidelberg: Heidelberger Akademie, 1959); cf. also Konrad Gaiser, *Platons ungeschreibene Lehre* (Stuttgart: E. Klept, 1963); Konrad Gaiser, ed., *Das Platonbild* (Hildesheim: G. Olms, 1969) and J. N. Findlay, *Plato: The Written and Unwritten Doctrines* (London: Routledge and Kegan Paul, 1974).

35. See Plato, *Gorgias* 503e-508c, for a reasonably clear and explicit statement of the nature of goodness (i.e., value).

36. G. W. v. Leibniz, "Monadology," no. 58 in G. R. Montgomery, trans., *Leibniz* (LaSalle, Ill.: Open Court, 1962), p. 263.

37. G. W. v. Leibniz, "Discourse on Metaphysics," sec. 6, in *Leibniz*, p. 11.

38. See, for example, Noel J. Brown, "Biological Diversity: The Global Challenge," in *Proceedings of the U.S. Strategy Conference on Biological Diversity* (see n. 11 above).

39. Leopold, *Sand County*, p. 224.

40. Peter Miller, "Value as Richness: Toward a Value Theory for an Expanded Naturalism in Environmental Ethics," *Environmental Ethics* 4 (1982): 103.

41. See Leibniz, "Monadology," nos. 53-59, for an explicit discussion of these conditions. In addition to order and variety, Leibniz also includes in the concept of value the tantalizingly "organic" characteristics of "interconnection," "relationship," "adaptation," and "universal harmony." For a fuller discussion see Walter H. O'Briant, "Leibniz's Contribution to Environmental Philosophy," *Environmental Ethics* 2 (1980): 215-20.

42. See Leibniz, "Discourse on Metaphysics," sec. 5, where he compares God to "an excellent Geometer" and to "a good architect." He goes on to say, "that the reason [God] wishes to avoid multiplicity of hypotheses or prin-

ciples [is] very much as the simplest system is preferred in Astronomy" (*Leibniz*, pp. 8-9).

43. Kenneth Goodpaster, "From Egoism to Environmentalism," in *Ethics and Problems of the 21st Century*, edited by Kenneth Goodpaster and Kenneth Sayre (Notre Dame: Notre Dame University Press, 1979), pp. 21-35.

44. Kant provides the clearest possible illustration: "Its [the categorical imperative's] foundation is this, that rational nature exists as an end in itself. Man necessarily conceives his own existence this way, and so far this is a subjective principle of human action." In Kant's view this subjective principle becomes (relatively) "objective" by generalization, viz.: "But in this way also every other rational being conceives of his own existence, and for the very same reason; hence the principle is also objective, and from it, as the highest practical ground, all laws of the will must be capable of being derived." John Stuart Mill, Bentham's utilitarian protégé, employs the same general strategy as Kant to transcend egoism. According to Mill, "the happiness [previously defined in terms of pleasure and pain] which forms the utilitarian standard of what is right in conduct is not the agent's own happiness but that of all concerned. As between *his own* happiness and that of others, utilitarianism requires him to be as strictly impartial as a disinterested and benevolent spectator." Immanuel Kant, *Foundations of the Metaphysics of Morals*, trans. by John Watson (Glasgow: Jackson, Wylie and Company, 1888), second section, and John Stuart Mill, *Utilitarianism* (New York: Bobbs-Merrill, 1957), chap. 2 (emphasis added).

45. Kant, *Foundations*, second section: "And even beings whose existence depends upon nature [including thus animals and plants], not upon our will have only relative value as means [i.e., instrumental value]. . . ." For a more elaborate statement see "Duties to Animals and Spirits," in Immanuel Kant, *Lectures on Ethics*, trans. by Louis Infield (New York: Harper and Row, 1963), pp. 239-41.

46. Jeremy Bentham, *An Introduction to the Principles of Morals and Legislation*, New Edition (Oxford: Oxford University Press, 1823), chap. xvii, sec. 1.

47. J. Baird Callicott, "Animal Liberation: A Triangular Affair," *Environmental Ethics* 2 (1980): 311-38. See also similar views expressed by R. and V. Routley, "Human Chauvinism and Environmental Ethics," in Mannison et al., eds., *Environmental Philosophy*, pp. 96-189 (see n. 12 above).

48. Peter Singer, "Not for Humans Only: The Place of Nonhumans in Environmental Issues," in Goodpaster and Sayre, eds., *Ethics and Problems of the 21st Century*, pp. 191-206. Tom Regan has expressed a similar view in *The Case for Animal Rights* (Berkeley: University of California Press, 1983), p. 360: ". . . the reason we ought to save the members of endangered species of animals is not because the species is endangered but because the individual animals have valid claims and thus rights. . . ."

49. In response to this concern which I first expressed as an "irony" of animal liberation in "A Triangular Affair," animal liberationist Edward Johnson

saw nothing wrong with it. According to Johnson, "the crucial point, though, is that there is no 'irony' here even if a species does become extinct, since it is not the species that is being liberated, but individual members of the species." Edward Johnson, "Animal Liberation Versus the Land Ethic," *Environmental Ethics* 3 (1981): 267.

50. Sagoff, personal communication.

51. Arthur Schopenhauer, *The World as Will and Idea*, trans. Haldane and Kemp (Garden City: Doubleday, 1961); see also, "Transcendent Considerations Concerning the Will as Thing in Itself," in *The Will to Live: Selected Writings of Arthur Schopenhauer*, edited by Richard Taylor (New York: Frederick Unger, 1962), pp. 33-42.

52. "Just as in my own will-to-live there is a yearning for more life . . . so the same obtains in all the will-to-live around me, equally whether it can express itself to my comprehension or whether it remains unvoiced." Schweitzer here says in effect, my essence and for me the source of my own preciousness is the will-to-live, but the same thing, a striving for life, is in every other living thing. There follows the transition from egoism to altruism: "Ethics thus consists in this, that I experience the necessity of practicing the same reverence for life toward all will-to-live, as toward my own." Albert Schweitzer, *Civilization and Ethics*, trans. John Naish, reprinted in Regan and Singer, eds., *Animal Rights and Human Obligations* (Englewood Cliffs: Prentice-Hall, 1976), p. 133.

53. H. J. McCloskey holds such a position, in "Rights," *Philosophical Quarterly* 15 (1965): 115-27. See also Meredith Williams, "Rights, Interests, and Moral Equality," *Environmental Ethics* (1980): 149-61. For the general relationship between interests and rights see also Joel Feinberg, "The Nature and Value of Rights," *Journal of Value Inquiry* 4 (1970): 243-57, and Bryan Norton, "Environmental Ethics and Non-human Rights," *Environmental Ethics* 4 (1982): 17-36.

54. Peter Singer, "All Animals Are Equal," in *Animal Rights*, p. 148, writes, "The capacity for suffering and enjoying things is a prerequisite for having interests at all, a condition that must be satisfied for having interests at all." See also Tom Regan, "The Moral Basis of Vegetarianism," *Canadian Journal of Philosophy* 5 (1975): 181-214; and William Frankena, "Ethics and the Environment," in Goodpaster and Sayre, eds., *Ethics of the 21st Century*.

55. Joel Feinberg, "Can Animals Have Rights?" in Regan and Singer, eds., *Animal Rights and Human Obligations*, p. 191. Paul W. Taylor, in "The Ethics of Respect for Nature," *Environmental Ethics* 3 (1981): 199-200, without using the term "conation," appears to understand "interests" along lines similar to Feinberg's: "We can act in a being's interest or contrary to its interest without its being interested in what we are doing to it. It may, indeed, be wholly unaware. . . . When construed in this way, the concept of a being's good [i.e., interest] is not coextensive with sentience or the capacity for feeling pain."

56. Kenneth Goodpaster, "On Being Morally Considerable," *Journal of Philos-*

ophy 75 (1978): 306-25. Goodpaster here wisely avoids a discussion of rights. In his view rights would involve something more (what more he does not say) than interests. J. Kantor, "The 'Interests' of Natural Objects," *Environmental Ethics* 2 (1980): 163-71, also draws attention to Feinberg's inconsistency in defining "interest" in terms of conation and then denying interests to plants. In Kantor's view plants may have interests. However, he does not think that the interests of plants can serve as the basis of rights; siding with Singer and Regan, he thinks that in addition a being must consciously suffer from having its interests harmed in order to be accorded rights.

57. Schweitzer, *Civilization and Ethics*, p. 136; Goodpaster, "Being Morally Considerable," p. 324. For a discussion of Schweitzer on this problem see William T. Blackstone, "The Search for an Environmental Ethic," in Tom Regan, ed., *Matters of Life and Death* (New York: Random House, 1980), pp. 299-335.

58. Goodpaster, "Being Morally Considerable," p. 313.

59. Schweitzer, *Civilization and Ethics*, p. 137.

60. Donald VanDeVeer, "Interspecific Justice," *Inquiry* 22 (1979): 55-79, has made an attempt to do just this.

61. Doubts concerning the serviceability of the predominant individual-egalitarian bias of moral metaphysics in the modern tradition vis-à-vis environmental ethical problems have been publicly expressed by John Rodman, "Liberation of Nature"; Bryan Norton, "Environmental Ethics and Non-Human Rights"; Richard and Val Routley, "Human Chauvinism"; Peter Miller, "Value as Richness"; Tom Regan, "The Nature and Possibility of an Environmental Ethic," *Environmental Ethics* 3 (1981): 19-34; J. Baird Callicott, "Animal Liberation."

62. Cf. David Hume, *A Treatise of Human Nature* (Oxford: The Clarendon Press, 1960), bk. III, pt. I.

63. Charles Darwin, *The Descent of Man and Selection in Relation to Sex*, second edition (New York: J. A. Hill, 1904), p. 97.

64. Ibid., p. 107.

65. Ibid., p. 118.

66. Ibid., p. 124.

67. Darwin seems both aware and forthright about his dependency on the concept of group selection in his account of the origin and evolution of morals: "We have now seen that actions are regarded by savages, and were probably so regarded by primeval man, as good or bad, solely as they obviously affect the welfare of the tribe,—not that of the species, nor that of the individual member of the species. This conclusion agrees well with the belief that the so-called moral sense is aboriginally derived from the social instincts, for both relate at first exclusively to the community" (ibid., p. 120). V. C. Wynne-Edwards, *Animal Dispersion in Relation to Social Behavior* (Edinburgh: Oliver and Boyd, 1962), provides the most celebrated recent support for group selection. Wynne-Edwards was refuted to the satisfaction at least of most biologists by G. C. Williams in *Adaptation and Natural Selection: A*

Critique of Some Current Evolutionary Thought (Princeton: Princeton University Press, 1966) and ever since, the concept of group selection has been avoided by most evolutionary theorists and certainly by sociobiologists. For a recent summary discussion see Michael Ruse, *Sociobiology: Sense or Nonsense?* (Boston: D. Reidel, 1979).

68. Most notably by W. D. Hamilton, "The Genetical Theory of Social Behavior," *Journal of Theoretical Biology* 7 (1964): 1-32; R. L. Trivers, "The Evolution of Reciprocal Altruism," *Quarterly Review of Biology* 46 (1971): 35-57; Edward O. Wilson, *Sociobiology: The New Synthesis* (Cambridge: Harvard University Press, 1975). Ruse, *Sociobiology: Sense or Nonsense?* provides a thorough bibliography.

69. Leopold, *Sand County*, p. 202.

70. Ibid.

71. Ehrlich and Ehrlich, *Extinction*, p. 48.

72. Ibid., pp. 50-51.

73. Hume, *Treatise*, pp. 484-85.

74. Darwin, *Descent*, p. 122.

75. Leopold, *Sand County*, p. 204.

76. See the discussion of Hume in W. D. Hudson, *Modern Moral Philosophy* (Garden City: Anchor Books, 1970), for a good summary of professional opinion.

7

Philosophical Problems for Environmentalism

ELLIOTT SOBER

I. Introduction

A number of philosophers have recognized that the environmental movement, whatever its practical political effectiveness, faces considerable theoretical difficulties in justification.[1] It has been recognized that traditional moral theories do not provide natural underpinnings for policy objectives and this has led some to skepticism about the claims of environmentalists, and others to the view that a revolutionary reassessment of ethical norms is needed. In this chapter, I will try to summarize the difficulties that confront a philosophical defense of environmentalism. I also will suggest a way of making sense of some environmental concerns that does not require the wholesale jettisoning of certain familiar moral judgments.

Preserving an endangered species or ecosystem poses no special conceptual problem when the instrumental value of that species or ecosystem is known. When we have reason to think that some natural object represents a resource to us, we obviously ought to take that fact into account in deciding what to do. A variety of potential uses may be under discussion, including food supply, medical applications, recreational use, and so on. As with any complex decision, it may be difficult even to agree on how to compare the competing values that may be involved. Willingness to pay in dollars is a familiar least common denominator, although it poses a number of problems. But here we have nothing that is specifically a problem for environmentalism.

The problem for environmentalism stems from the idea that species and ecosystems ought to be preserved for reasons additional to their known value as resources for human use. The feeling is that even when we cannot say what nutritional, medicinal, or recreational benefit the preservation provides, there still is a value in preservation. It is the search for a rationale for this feeling that constitutes the main conceptual problem for environmentalism.

The problem is especially difficult in view of the holistic (as opposed to individualistic) character of the things being assigned value. Put simply, what is special about environmentalism is that it values the preservation of species, communities, or ecosystems, rather than the individual organisms of which they are composed. "Animal liberationists" have urged that we should take the suffering of sentient animals into account in ethical deliberation.[2] Such beasts are not mere things to be used as cruelly as we like no matter how trivial the benefit we derive. But in "widening the ethical circle," we are simply including in the community more individual organisms whose costs and benefits we compare. Animal liberationists are extending an old and familiar ethical doctrine—namely, utilitarianism—to take account of the welfare of other individuals. Although the practical consequences of this point of view may be revolutionary, the theoretical perspective is not at all novel. If suffering is bad, then it is bad for any individual who suffers.[3] Animal liberationists merely remind us of the consequences of familiar principles.[4]

But trees, mountains, and salt marshes do not suffer. They do not experience pleasure and pain, because, evidently, they do not have experiences at all. The same is true of species. Granted, individual organisms may have mental states; but the species—taken to be a population of organisms connected by certain sorts of interactions (preeminently, that of exchanging genetic material in reproduction)—does not. Or put more carefully, we might say that the only sense in which species have experiences is that their member organisms do: the attribution at the population level, if true, is true simply in virtue of its being true at the individual level. Here is a case where reductionism is correct.

So perhaps it is true in this reductive sense that some species experience pain. But the values that environmentalists attach to preserving species do not reduce to any value of preserving organisms. It is in this sense that environmentalists espouse a holistic value system. Environmentalists care about entities that by no stretch of the imagination have experiences (e.g., mountains). What is more, their position does not force them to care if individual organisms suffer pain, so long as the species is preserved. Steel traps may outrage an animal liberationist because of the suffering they inflict, but an environmentalist aiming just at the preservation of a balanced ecosystem might see here no cause for complaint. Similarly, environmentalists think that the distinction between wild and domesticated organisms is important, in that it is the preservation of "natural" (i.e., not created by the "artificial interference" of human beings) objects that matters, whereas animal liberationists see the main problem in terms of the suffering of any organism—domesticated or not.[5] And finally, environmentalists and animal liberationists diverge on what might be called the $n + m$ question. If two species—say blue and sperm whales—have roughly comparable capacities for experiencing pain, an animal liberationist might tend to think of the preservation of a sperm whale as wholly on an ethical par with the preservation of a blue whale. The fact that one organism is part of an endangered species while the other is not does not make the rare individual more intrinsically important. But for an environmentalist, this holistic property—membership in an endangered species—makes all the difference in the world: a world with n sperm and m blue whales is far better than a world with $n + m$ sperm and 0 blue whales. Here we have a stark contrast between an ethic in which it is the life situation of individuals that matters, and an ethic in which the stability and diversity of populations of individuals are what matter.[6]

Both animal liberationists and environmentalists wish to broaden our ethical horizons—to make us realize that it is not just human welfare that counts. But they do this in very different, often conflicting, ways. It is no accident that at the level of practical politics the two points of view increasingly find themselves at loggerheads.[7] This practical conflict is the expression of a deep theoretical divide.

II. THE IGNORANCE ARGUMENT

"Although we might not now know what use a particular endangered species might be to us, allowing it to go extinct forever closes off the possibility of discovering and exploiting a future use." According to this point of view, our ignorance of value is turned into a reason for action. The scenario envisaged in this environmentalist argument is not without precedent; who could have guessed that penicillin would be good for something other than turning out cheese? But there is a fatal defect in such arguments, which we might summarize with the phrase *out of nothing, nothing comes*: rational decisions require assumptions about what is true and what is valuable (in decision-theoretic jargon, the inputs must be probabilities and utilities). If you are completely ignorant of values, then you are incapable of making a rational decision, either for or against preserving some species. The fact that you do not know the value of a species, by itself, cannot count as a reason for wanting one thing rather than another to happen to it.

And there are so many species. How many geese that lay golden eggs are there apt to be in that number? It is hard to assign probabilities and utilities precisely here, but an analogy will perhaps reveal the problem confronting this environmentalist argument. Most of us willingly fly on airplanes, when safer (but less convenient) alternative forms of transportation are available. Is this rational? Suppose it were argued that there is a small probability that the next flight you take will crash. This would be very bad for you. Is it not crazy for you to risk this, given that the only gain to you is that you can reduce your travel time by a few hours (by not going by train, say)? Those of us who not only fly, but congratulate ourselves for being rational in doing so, reject this argument. We are prepared to accept a small chance of a great disaster in return for the high probability of a rather modest benefit. If this is rational, no wonder that we might consistently be willing to allow a species to go extinct in order to build a hydroelectric plant.

That the argument from ignorance is no argument at all can be seen from another angle. If we literally do not know what consequences the extinction of this or that species may bring, then we should take seriously the possibility that the extinction may be beneficial as well as the possibility that it may be deleterious. It may sound deep to insist that we preserve endangered species precisely because we do not know why they are valuable. But ignorance on a scale like this cannot provide the basis for any rational action.

Rather than invoke some unspecified future benefit, an environmentalist may argue that the species in question plays a crucial role in stabilizing the ecosystem of which it is a part. This will undoubtedly be true for carefully chosen species and ecosystems, but one should not generalize this argument into a global claim to the effect that *every* species is crucial to a balanced ecosystem. Although ecologists used to agree that the complexity of an ecosystem stabilizes it, this hypothesis has been subject to a number of criticisms and qualifications, both from a theoretical and an empirical perspective.[8] And for certain kinds of species (those which occupy a rather small area and whose normal population is small) we can argue that extinction would probably not disrupt the community. However fragile the biosphere may be, the extreme view that everything is crucial is almost certainly not true.

But, of course, environmentalists are often concerned by the fact that extinctions are occurring now at a rate much higher than in earlier times. It is mass extinction that threatens the biosphere, they say, and this claim avoids the spurious assertion that communities are so fragile that even one extinction will cause a crash. However, if the point is to avoid a mass extinction of species, how does this provide a rationale for preserving a species of the kind just described, of which we rationally believe that its passing will not destabilize the ecosystem? And, more generally, if mass extinction is known to be a danger to us, how does this translate into a value for preserving any particular species? Notice that we have now passed beyond the confines of the argument from ignorance; we are taking as a premise the idea that mass extinction would be a catastrophe (since it would destroy the ecosystem on which we depend). But how should that premise affect our valuing the California condor, the blue whale, or the snail darter?

III. THE SLIPPERY SLOPE ARGUMENT

Environmentalists sometimes find themselves asked to explain why each species matters so much to them, when there are, after all, so many. We may know of special reasons for valuing particular species, but how can we justify thinking that each and every species is important? "Each extinction impoverishes the biosphere" is often the answer given, but it really fails to resolve the issue. Granted, each extinction impoverishes, but it only impoverishes a little bit. So if it is the *wholesale* impoverishment of the biosphere that matters, one would apparently have to concede that each extinction matters a little, but only a little. But environmentalists may be loathe to concede this, for if they concede that each species matters only a little, they seem to be inviting the wholesale impoverishment that would be an unambiguous disaster.[9] So they dig in their heels and insist that each species matters a lot. But to take this line, one must find some other rationale than the idea that mass extinction would be a great harm. Some of these alternative rationales we will examine later. For now, let us take a closer look at the train of thought involved here.

. Slippery slopes are curious things: if you take even one step onto them, you inevitably slide all the way to the bottom. So if you want to avoid finding yourself at the bottom, you must avoid stepping onto them at all. To mix metaphors, stepping onto a slippery slope is to invite being nickeled and dimed to death.

Slippery slope arguments have played a powerful role in a number of recent ethical debates. One often hears people defend the legitimacy of abortions by arguing that since it is permissible to abort a single-celled fertilized egg, it must be permissible to abort a foetus of any age, since there is no place to draw the line from 0 to 9 months. Antiabortionists, on the other hand, sometimes argue in the other direction: since infanticide of newborns is not permissible, abortion at any earlier time is also not allowed, since there is no place to draw the line. Although these two arguments reach opposite conclusions about the permissibility of abortions, they agree on the following idea: since there is no principled place to draw the line on the continuum from newly fertilized egg to foetus gone to term, one must treat all these cases in the same way. Either abortion is always permitted or it never is, since there is no place to draw the line. Both sides run their favorite slippery slope arguments, but try to precipitate slides in opposite directions.

Starting with 10 million extant species, and valuing overall diversity, the environmentalist does not want to grant that each species matters only a little. For having granted this, commercial expansion and other causes will reduce the tally to 9,999,999. And then the argument is repeated, with each species valued only a little, and diversity declines another notch. And so we are well on our way to a considerably impoverished biosphere, a little at a time. Better to reject the starting premise—namely, that each species matters only a little—so that the slippery slope can be avoided.

Slippery slopes should hold no terror for environmentalists, because it is often a mistake to demand that a line be drawn. Let me illustrate by an example. What is the difference between being bald and not? Presumably, the difference concerns the number of hairs you have on your head. But what is the precise number of hairs marking the boundary between baldness and not being bald? There is no such number. Yet, it would be a fallacy to conclude that there is no difference between baldness and hairiness. The fact that you cannot draw a line does not force you to say that the two alleged categories collapse into one. In the abortion case, this means that even if there is no precise point in foetal development that involves some discontinuous, qualitative change, one is still not obliged to think of newly fertilized eggs and foetuses gone to term as morally on a par. Since the biological differences are ones of degree, not kind, one may want to adopt the position that the moral differences are likewise matters of degree. This may lead to the view that a woman should have a better reason for having an abortion, the more developed her foetus is. Of course, this position does not logically follow from the idea that there is no place to draw the line; my point is just that differences in degree do not demolish the possibility of there being real moral differences.

In the environmental case, if one places a value on diversity, then each species becomes more valuable as the overall diversity declines. If we begin with 10 million species, each may matter little, but as extinctions continue, the remaining ones matter more and more. According to this outlook, a better and better reason would be demanded for allowing yet another species to go extinct. Perhaps certain sorts of economic development would justify the extinction of a species at one time. But granting this does not oblige one to conclude that the same sort of decision would have to be made further down the road. This means that one can value diversity without being obliged to take the somewhat exaggerated position that each species, no matter how many there are, is terribly precious in virtue of its contribution to that diversity.

Yet, one can understand that environmentalists might be reluctant to concede this point. They may fear that if one now allows that most species contribute only a little to overall diversity, one will set in motion a political process that cannot correct itself later. The worry is that even when the overall diversity has been drastically reduced, our ecological sensitivities will have been so coarsened that we will no longer be in a position to realize (or to implement policies fostering) the preciousness of what is left. This fear may be quite justified, but it is important to realize that it does not conflict with what was argued above. The political utility of making an argument should not be confused with the argument's soundness.

The fact that you are on a slippery slope, by itself, does not tell you whether you are near the beginning, in the middle, or at the end. If species diversity is a matter of degree, where do we currently find ourselves— on the verge of catastrophe, well on our way in that direction, or at some distance from a global crash? Environmentalists often urge that we are fast approaching a precipice; if we are, then the reduction in diversity that every succeeding extinction engenders should be all we need to justify species preservation.[10]

Sometimes, however, environmentalists advance a kind of argument not predicated on the idea of fast approaching doom. The goal is to show that there is something wrong with allowing a species to go extinct (or with causing it to go extinct), even if overall diversity is not affected much. I now turn to one argument of this kind.

IV. Appeals to What Is Natural

I noted earlier that environmentalists and animal liberationists disagree over the significance of the distinction between wild and domesticated animals. Since both types of organisms can experience pain, animal liberationists will think of each as meriting ethical consideration. But environmentalists will typically not put wild and domesticated organisms on a par.[11] Environmentalists typically are interested in preserving what is natural, be it a species living in the wild or a wilderness ecosystem. If a kind of domesticated chicken were threatened with extinction, I doubt that environmental groups would be up in arms. And if certain unique types of human environments—say urban slums in the United States—

were "endangered," it is similarly unlikely that environmentalists would view this process as a deplorable impoverishment of the biosphere.

The environmentalist's lack of concern for humanly created organisms and environments may be practical rather than principled. It may be that at the level of values, no such bifurcation is legitimate, but that from the point of view of practical political action, it makes sense to put one's energies into saving items that exist in the wild. This subject has not been discussed much in the literature, so it is hard to tell. But I sense that the distinction between wild and domesticated has a certain theoretical importance to many environmentalists. They perhaps think that the difference is that we created domesticated organisms which would otherwise not exist, and so are entitled to use them solely for our own interests. But we did not create wild organisms and environments, so it is the height of presumption to expropriate them for our benefit. A more fitting posture would be one of "stewardship": we have come on the scene and found a treasure not of our making. Given this, we ought to preserve this treasure in its natural state.

I do not wish to contest the appropriateness of "stewardship." It is the dichotomy between artificial (domesticated) and natural (wild) that strikes me as wrong-headed. I want to suggest that to the degree that "natural" means anything biologically, it means very little ethically. And, conversely, to the degree that "natural" is understood as a normative concept, it has very little to do with biology.

Environmentalists often express regret that we human beings find it so hard to remember that we are part of nature—one species among many others—rather than something standing outside of nature. I will not consider here whether this attitude is cause for complaint; the important point is that seeing us as part of nature rules out the environmentalist's use of the distinction between artificial-domesticated and natural-wild described above. *If we are part of nature, then everything we do is part of nature, and is natural in that primary sense.*[12] When we domesticate organisms and bring them into a state of dependence on us, this is simply an example of one species exerting a selection pressure on another. If one calls this "unnatural," one might just as well say the same of parasitism or symbiosis (compare human domestication of animals and plants and "slave-making" in the social insects).

The concept of naturalness is subject to the same abuses as the concept of normalcy. *Normal* can mean *usual* or it can mean *desirable*. Although only the total pessimist will think that the two concepts are mutually exclusive, it is generally recognized that the mere fact that something is common does not by itself count as a reason for thinking that it is desirable. This distinction is quite familiar now in popular discussions of mental health, for example. Yet, when it comes to environmental issues, the concept of naturalness continues to live a double life. The destruction of wilderness areas by increased industrialization is bad because it is unnatural. And it is unnatural because it involves transforming a natural into an artificial habitat. Or one might hear that although extinction is a natural process, the kind of mass extinction currently being precipitated by our species is unprecedented, and so is unnatural. Environmentalists should look elsewhere for a defense of their policies, lest conservation simply become a variant of uncritical conservatism in which the axiom "Whatever is, is right" is modified to read "Whatever is (before human beings come on the scene), is right."

This conflation of the biological with the normative sense of "natural" sometimes comes to the fore when environmentalists attack animal liberationists for naive do-goodism. Callicott writes:

> . . . the value commitments of the humane movement seem at bottom to betray a world-denying or rather a life-loathing philosophy. The natural world as actually constituted is one in which one being lives at the expense of others. Each organism, in Darwin's metaphor, struggles to maintain its own organic integrity. . . . To live *is* to be anxious about life, to feel pain and pleasure in a fitting mixture, and sooner or later to die. That is the way the system works. *If nature as a whole is good, then pain and death are also good.* Environmental ethics in general require people to play fair in the natural system. The neo-Benthamites have in a sense taken the uncourageous approach. People have attempted to exempt themselves from the life/death reciprocities of natural processes and from ecological limitations in the name of a prophylactic ethic of maximizing rewards (pleasure) and minimizing unwelcome information (pain). To be fair, the humane moralists seem to suggest that we should attempt to project the same values into the nonhuman animal world and to widen the charmed circle—no matter that it would be biologically unrealistic to do so or biologically ruinous if, per impossible, such an environmental ethic were implemented.
>
> There is another approach. Rather than imposing our alienation from nature and natural processes and cycles of life on other animals, we human beings could reaffirm our participation in nature by accepting life as it is given without a sugar coating. . . .[13]

On the same page, Callicott quotes with approval Shepard's remark that "the humanitarian's projection onto nature of illegal murder and the rights of civilized people to safety not only misses the point but is exactly contrary to fundamental ecological reality: the structure of nature is a sequence of killings."[14]

Thinking that what is found in nature is beyond ethical defect has not always been popular. Darwin wrote:

> . . . That there is much suffering in the world no one disputes.
>
> Some have attempted to explain this in reference to man by imagining that it serves for his moral improvement. But the number of men in the world is as nothing compared with that of all other sentient beings, and these often suffer greatly without any moral improvement. A being so powerful and so full of knowledge as a God who could create the universe, is to our finite minds omnipotent and omniscient, and it revolts our understanding to suppose that his benevolence is not unbounded, for what advantage can there be in the sufferings of millions of the lower animals throughout almost endless time? This very old argument from the existence of suffering against the existence of an intelligent first cause seems to me a strong one; whereas, as just remarked, the presence of much suffering agrees well with the view that all organic beings have been developed through variation and natural selection.[15]

Darwin apparently viewed the quantity of pain found in nature as a melancholy and sobering consequence of the struggle for existence. But once we adopt the Panglossian attitude that this is the best of all possible worlds ("there is just the right amount of pain," etc.), a failure to identify what is natural with what is good can only seem "world-denying," "life-loathing," "in a sense uncourageous," and "contrary to fundamental ecological reality."[16]

Earlier in his essay, Callicott expresses distress that animal liberationists fail to draw a sharp distinction "between the very different plights (and rights) of wild and domestic animals."[17] Domestic animals are creations of man, he says. "They are living artifacts, but artifacts nevertheless. . . . There is thus something profoundly incoherent (and insensitive as well) in the complaint of some animal liberationists that the 'natural behavior' of chickens and bobby calves is cruelly frustrated on factory farms. It would make almost as much sense to speak of the natural behavior of tables and chairs."[18] Here again we see teleology playing a decisive role: wild organisms do not have the natural function of serving human ends, but domesticated animals do. Cheetahs in zoos are crimes against what is natural; veal calves in boxes are not.

The idea of "natural tendency" played a decisive role in pre-Darwinian biological thinking. Aristotle's entire science—both his physics and his biology—is articulated in terms of specifying the natural tendencies of kinds of objects and the interfering forces that can prevent an object from achieving its intended state.[19] Heavy objects in the sublunar sphere have location at the center of the earth as their natural state; each tends to go there, but is prevented from doing so.[20] Organisms likewise are conceptualized in terms of this natural state model:

> ... [for] any living thing that has reached its normal development and which is unmutilated, and whose mode of generation is not spontaneous, the most natural act is the production of another like itself, an animal producing an animal, a plant a plant. . . .[21]

But many interfering forces are possible, and in fact the occurrence of "monsters" is anything but uncommon. According to Aristotle, mules (sterile hybrids) count as deviations from the natural state. In fact, females are monsters as well, since the natural tendency of sexual reproduction is for the offspring to perfectly resemble the father, who, according to Aristotle, provides the "genetic instructions" (to put the idea anachronistically) while the female provides only the matter.[22]

What has happened to the natural state model in modern science? In physics, the idea of describing what a class of objects will do in the absence of "interference" lives on: Newton specified this "zero-force state" as rest or uniform motion, and in general relativity, this state is understood in terms of motion along geodesics. But one of the most profound achievements of Darwinian biology has been the jettisoning of this kind of model.[23] It isn't just that Aristotle was wrong in his detailed claims about mules and women; the whole structure of the natural state model has been discarded. Population biology is not conceptualized in terms of positing some characteristic that all members of a species would have in common, were interfering forces absent. Variation is not thought of as a deflection from the natural state of uniformity. Rather, variation is taken to be a fundamental property in its own right. Nor, at the level of individual biology, does the natural state model find an application. Developmental theory is not articulated by specifying a natural tendency and a set of interfering forces. The main conceptual tool for describing the various developmental pathways open to a genotype is the norm of reaction.[24] The norm of reaction of a genotype within a range of environments will describe what phenotype the genotype will produce in a given environment. Thus, the norm of reaction for a corn plant genotype might describe how its height is influenced by the amount of moisture in the soil. The norm of reaction is entirely silent on which phenotype is the "natural" one. The idea that a corn plant might have some "natural height," which can be augmented or diminished by "interfering forces" is entirely alien to post-Darwinian biology.

The fact that the concepts of natural state and interfering force have lapsed from biological thought does not prevent environmentalists from inventing them anew. Perhaps these concepts can be provided with some sort of normative content; after all, the normative idea of "human rights" may make sense even if it is not a theoretical underpinning of any empirical science. But environmentalists should not assume that they can rely on some previously articulated scientific conception of "natural."

V. Appeals to Needs and Interests

The version of utilitarianism considered earlier (according to which something merits ethical consideration if it can experience pleasure and/or pain) leaves the environmentalist in the lurch. But there is an alternative to Bentham's hedonistic utilitarianism that has been thought by some to be a foundation for environmentalism. Preference utilitarianism says that an object's having interests, needs, or preferences gives it ethical status. This doctrine is at the core of Stone's affirmative answer to the title question of his book *Should Trees Have Standing?*[25] "Natural objects *can* communicate their wants (needs) to us, and in ways that are not terribly ambiguous. . . . The lawn tells me that it wants water by a certain dryness of the blades and soil—immediately obvious to the touch—the appearance of bald spots, yellowing, and a lack of springiness after being walked on." And if plants can do this, presumably so can mountain ranges, and endangered species. Preference utilitarianism may thereby seem to grant intrinsic ethical importance to precisely the sorts of objects about which environmentalists have expressed concern.

The problems with this perspective have been detailed by Sagoff.[26] If one does not require of an object that it have a mind for it to have wants or needs, what *is* required for the possession of these ethically relevant properties? Suppose one says that an object needs something if it will cease to exist if it does not get it. Then species, plants, and mountain ranges have needs, but only in the sense that automobiles, garbage dumps, and buildings do too. If everything has needs, the advice to take needs into account in ethical deliberation is empty, unless it is supplemented by some technique for weighting and comparing the needs of different objects. A corporation will go bankrupt unless a highway is built. But the swamp will cease to exist if the highway is built. Perhaps one should take into account all relevant needs, but the question is how to do this in the event that needs conflict.

Although the concept of need can be provided with a permissive, all-inclusive definition, it is less easy to see how to do this with the concept of want. Why think that a mountain range "wants" to retain its unspoiled appearance, rather than house a new amusement park?[27] Needs are not at issue here, since in either case, the mountain continues to exist. One might be tempted to think that natural objects like mountains and species have "natural tendencies," and that the concept of want should be liberalized so as to mean that natural objects "want" to persist in their natural states. This Aristotelian view, as I argued in the previous section, simply makes no sense.[28] Granted, a commercially undeveloped mountain will persist in this state, unless it is commercially developed. But it is equally true that a commercially untouched hill will become commercially developed, unless something causes this not to happen. I see no hope for extending the concept of wants to the full range of objects valued by environmentalists.

The same problems emerge when we try to apply the concepts of needs and wants to species. A species may need various resources, in the sense that these are necessary for its continued existence. But what do species want? Do they want to remain stable in numbers, neither growing nor shrinking? Or since most species have gone extinct, perhaps what species really want is to go extinct, and it is human meddlesomeness that frustrates this natural tendency? Preference utilitarianism is no more likely than hedonistic utilitarianism to secure autonomous ethical status for endangered species.

Ehrenfeld describes a related distortion that has been inflicted on the diversity/stability hypothesis in theoretical ecology.[29] If it were true that increasing the diversity of an ecosystem causes it to be more stable, this might encourage the Aristotelian idea that ecosystems have a natural tendency to increase their diversity. The full realization of this tendency— the natural state that is the goal of ecosystems—is the "climax" or "mature" community. Extinction diminishes diversity, so it frustrates ecosystems from attaining their goal. Since the hypothesis that diversity causes stability is now considered controversial (to say the least), this line of thinking will not be very tempting. But even if the diversity/stability hypothesis were true, it would not permit the environmentalist to conclude that ecosystems have an interest in retaining their diversity.

Darwinism has not banished the idea that parts of the natural world are goal-directed systems, but has furnished this idea with a natural mechanism. We properly conceive of organisms (or genes, sometimes) as being in the business of maximizing their chances of survival and reproduction. We describe characteristics as adaptations—as devices that exist for the furtherance of these ends. Natural selection makes this perspective intelligible. But Darwinism is a profoundly individualistic doctrine.[30] Darwinism rejects the idea that species, communities, and ecosystems have adaptations that exist for their own benefit. These higher-level entities

185

are not conceptualized as goal-directed systems; what properties of organization they possess are viewed as artifacts of processes operating at lower levels of organization. An environmentalism based on the idea that the ecosystem is directed toward stability and diversity must find its foundation elsewhere.

VI. Granting Wholes Autonomous Value

A number of environmentalists have asserted that environmental values cannot be grounded in values based on regard for individual welfare. Aldo Leopold wrote in *A Sand County Almanac* that "a thing is right when it tends to preserve the integrity, stability, and beauty of the biotic community. It is wrong when it tends otherwise."[31] Callicott develops this idea at some length, and ascribes to ethical environmentalism the view that "the preciousness of individual deer, *as of any other specimen*, is inversely proportional to the population of the species."[32] In his *Desert Solitaire*, Edward Abbey notes that he would sooner shoot a man than a snake.[33] And Garrett Hardin asserts that human beings injured in wilderness areas ought not to be rescued: making great and spectacular efforts to save the life of an individual "makes sense only when there is a shortage of people. I have not lately heard that there is a shortage of people."[34] The point of view suggested by these quotations is quite clear. It isn't that preserving the integrity of ecosystems has autonomous value, to be taken into account just as the quite distinct value of individual human welfare is. Rather, the idea is that the only value is the holistic one of maintaining ecological balance and diversity. Here we have a view that is just as monolithic as the most single-minded individualism; the difference is that the unit of value is thought to exist at a higher level of organization.

It is hard to know what to say to someone who would save a mosquito, just because it is rare, rather than a human being, if there were a choice. In ethics, as in any other subject, rationally persuading another person requires the existence of shared assumptions. If this monolithic environmentalist view is based on the notion that ecosystems have needs and interests, and that these take total precedence over the rights and interests of individual human beings, then the discussion of the previous sections is relevant. And even supposing that these higher-level entities have needs and wants, what reason is there to suppose that these matter and that the wants and needs of individuals matter not at all? But if this source of defense is jettisoned, and it is merely asserted that only ecosystems have value, with no substantive defense being offered, one must begin by requesting an argument: *why* is ecosystem stability and diversity the only value?

Some environmentalists have seen the individualist bias of utilitarianism as being harmful in ways additional to its impact on our perception of ecological values. Thus, Callicott writes:

> On the level of social organization, the interests of society may not always coincide with the sum of the interests of its parts. Discipline, sacrifice, and individual restraint are often necessary in the social sphere to maintain social integrity as within the bodily organism. A society, indeed, is particularly vulnerable to disintegration when its members become preoccupied totally with their own particular interest, and ignore those distinct and independent interests of the community as a whole. One example, unfortunately, our own society, is altogether too close at hand to be examined with strict academic detachment. The United States seems to pursue uncritically a social policy of reductive utilitarianism, aimed at promoting the happiness of all its members severally. Each special interest accordingly clamors more loudly to be satisfied while the community as a whole becomes noticeably more and more infirm economically, environmentally, and politically.[35]

Callicott apparently sees the emergence of individualism and alienation from nature as two aspects of the same process. He values "the symbiotic relationship of Stone Age man to the natural environment" and regrets that "civilization has insulated and alienated us from the rigors and challenges of the natural environment. The hidden agenda of the humane ethic," he says, "is the imposition of the anti-natural prophylactic ethos of comfort and soft pleasure on an even wider scale. The land ethic, on the other hand, requires a shrinkage, if at all possible, of the domestic sphere; it rejoices in a recrudescence of the wilderness and a renaissance of tribal cultural experience."[36]

Callicott is right that "strict academic detachment" is difficult here. The reader will have to decide whether the United States currently suffers from too much or too little regard "for the happiness of all its members severally" and whether we should feel nostalgia or pity in contemplating what the Stone Age experience of nature was like.

VII. THE DEMARCATION PROBLEM

Perhaps the most fundamental theoretical problem confronting an environmentalist who wishes to claim that species and ecosystems have autonomous value is what I will call the *problem of demarcation*. Every ethical theory must provide principles that describe which objects matter for their own sakes and which do not. Besides marking the boundary between these two classes by enumerating a set of ethically relevant properties, an ethical theory must say why the properties named, rather than others, are the ones that count. Thus, for example, hedonistic utilitarianism cites the capacity to experience pleasure and/or pain as the decisive criterion; preference utilitarianism cites the having of preferences (or wants, or interests) as the decisive property. And a Kantian ethical theory will include an individual in the ethical community only if it is capable of rational reflection and autonomy.[37] Not that justifying these various proposed solutions to the demarcation problem is easy; indeed, since this issue is so fundamental, it will be very difficult to justify one proposal as opposed to another. Still, a substantive ethical theory is obliged to try.

Environmentalists, wishing to avoid the allegedly distorting perspective of individualism, frequently want to claim autonomous value for wholes. This may take the form of a monolithic doctrine according to which the only thing that matters is the stability of the ecosystem. Or it may embody a pluralistic outlook according to which ecosystem stability and species preservation have an importance additional to the welfare of individual organisms. But an environmentalist theory shares with all ethical theories an interest in not saying that everything has autonomous value. The reason this position is proscribed is that it makes the adjudication of ethical conflict very difficult indeed. (In addition, it is radically implausible, but we can set that objection to one side.)

Environmentalists, as we have seen, may think of natural objects, like mountains, species, and ecosystems, as mattering for their own sake, but of artificial objects, like highway systems and domesticated animals, as having only instrumental value. If a mountain and a highway are both made of rock, it seems unlikely that the difference between them arises from the fact that mountains have wants, interests, and preferences, but highway systems do not. But perhaps the place to look for the relevant difference is not in their present physical composition, but in the historical fact of how each came into existence. Mountains were created by natural processes, whereas highways are humanly constructed. But once we realize that organisms construct their environments in nature, this contrast begins to cloud.[38] Organisms do not passively reside in an environment whose properties are independently determined. Organisms transform their environments by physically interacting with them. An anthill is an artifact just as a highway is. Granted, a difference obtains at the level of whether conscious deliberation played a role, but can one take seriously the view that artifacts produced by conscious planning are thereby *less* valuable than ones that arise without the intervention of mentality?[39] As we have noted before, although environmentalists often accuse their critics of failing to think in a biologically realistic way, their use of the distinction between "natural" and "artificial" is just the sort of idea that stands in need of a more realistic biological perspective.

My suspicion is that the distinction between natural and artificial is not the crucial one. On the contrary, certain features of environmental concerns imply that natural objects are exactly on a par with certain artificial ones. Here the intended comparison is not between mountains and highways, but between mountains and works of art. My goal in what follows is not to sketch a substantive conception of what determines the value of objects in these two domains, but to motivate an analogy.

For both natural objects and works of art, our values extend beyond the concerns we have for experiencing pleasure. Most of us value seeing an original painting more than we value seeing a copy, even when we could not tell the difference. When we experience works of art, often what we value is not just the kinds of experiences we have, but, in addition, the connections we usually have with certain real objects. Routley and Routley have made an analogous point about valuing the wilderness experience: a "wilderness experience machine" that caused certain sorts of hallucinations would be no substitute for actually going into the wild.[40] Nor is this fact about our valuation limited to such aesthetic and environmentalist contexts. We love various people in our lives. If a molecule-for-molecule replica of a beloved person were created, you would not love that individual, but would continue to love the individual to whom you actually were historically related.[41] Here again, our attachments are to objects and people as they really are, and not just to the experiences that they facilitate.

Another parallel between environmentalist concerns and aesthetic values concerns the issue of context. Although environmentalists often stress the importance of preserving endangered species, they would not be completely satisfied if an endangered species were preserved by putting a number of specimens in a zoo or in a humanly constructed preserve. What is taken to be important is preserving the species in its natural habitat. This leads to the more holistic position that preserving ecosystems, and not simply preserving certain member species, is of primary importance. Aesthetic concerns often lead in the same direction. It was not merely saving a fresco or an altar piece that motivated art historians after the most recent flood in Florence. Rather, they wanted to save these works of art in their original ("natural") settings. Not just the painting, but the church that housed it; not just the church, but the city itself. The idea of objects residing in a "fitting" environment plays a powerful role in both domains.

Environmentalism and aesthetics both see value in rarity. Of two whales, why should one be more worthy of aid than another, just because one belongs to an endangered species? Here we have the $n + m$ question mentioned in Section I. As an ethical concern, rarity is difficult to understand. Perhaps this is because our ethical ideas concerning justice and equity (note the word) are saturated with individualism. But in the context of aesthetics, the concept of rarity is far from alien. A work of art may have enhanced value simply because there are very few other works by the same artist, or from the same historical period, or in the same style. It isn't that the price of the item may go up with rarity; I am talking about aesthetic value, not monetary worth. Viewed as valuable aesthetic objects, rare organisms may be valuable because they are rare.

A disanalogy may suggest itself. It may be objected that works of art are of instrumental value only, but that species and ecosystems have intrinsic value. Perhaps it is true, as claimed before, that our attachment to works of art, to nature, and to our loved ones extends beyond the experiences they allow us to have. But it may be argued that what is valuable in the aesthetic case is always the relation of a valuer to a valued object.[42] When we experience a work of art, the value is not simply in the experience, but in the composite fact that we and the work of art are related in certain ways. This immediately suggests that if there were no valuers in the world, nothing would have value, since such relational facts could no longer obtain. So, to adapt Routley and Routley's "last man argument," it would seem that if an ecological crisis precipitated a collapse of the world system, the last human being (whom we may assume for the purposes of this example to be the last valuer) could set about destroying all works of art, and there would be nothing wrong in this.[43] That is, if aesthetic objects are valuable only in so far as valuers can stand in certain relations to them, then when valuers disappear, so does the possibility of aesthetic value. This would deny, in one sense, that aesthetic objects are intrinsically valuable: it isn't they, in themselves, but rather the relational facts that they are part of, that are valuable.

In contrast, it has been claimed that the "last man" would be wrong to destroy natural objects such as mountains, salt marshes, and species.[44] (So as to avoid confusing the issue by bringing in the welfare of individual organisms, Routley and Routley imagine that destruction and mass extinctions can be caused painlessly, so that there would be nothing wrong about this undertaking from the point of view of the nonhuman organisms involved.) If the last man ought to preserve these natural objects, then these objects appear to have a kind of autonomous value; their value would extend beyond their possible relations to valuers. If all this were true, we would have here a contrast between aesthetic and natural objects, one that implies that natural objects are more valuable than works of art.

Routley and Routley advance the last man argument as if it were decisive in showing that environmental objects such as mountains and salt marshes have autonomous value. I find the example more puzzling than decisive. But, in the present context, we do not have to decide whether Routley and Routley are right. We only have to decide whether this imagined situation brings out any relevant difference between aesthetic and environmental values. Were the last man to look up on a certain hillside, he would see a striking rock formation next to the ruins of a Greek temple. Long ago the temple was built from some of the very rocks that still stud the slope. Both promontory and temple have a history, and both have been transformed by the biotic and the abiotic environments. I myself find it impossible to advise the last man that the peak matters more than the temple. I do not see a relevant difference. Environmentalists, if they hold that the solution to the problem of demarcation is to be found in the distinction between natural and artificial, will have to find such a distinction. But if environmental values are aesthetic, no difference need be discovered.

Environmentalists may be reluctant to classify their concern as aesthetic. Perhaps they will feel that aesthetic concerns are frivolous. Perhaps they will feel that the aesthetic regard for artifacts that has been made possible by culture is antithetical to a proper regard for wilderness. But such contrasts are illusory. Concern for environmental values does not require a stripping away of the perspective afforded by civilization; to value the wild, one does not have to "become wild" oneself (whatever that may mean). Rather, it is the material comforts of civilization that make possible a serious concern for both aesthetic and environmental values. These are concerns that can become pressing in developed nations in part because the populations of those countries now enjoy a certain substantial level of prosperity. It would be the height of condescension to expect a nation experiencing hunger and chronic disease to be inordinately concerned with the autonomous value of ecosystems or with creating and preserving works of art. Such values are not frivolous, but they can become important to us only after certain fundamental human needs are satisfied. Instead of radically jettisoning individualist ethics, environmentalists may find a more hospitable home for their values in a category of value that has existed all along.[45]

NOTES

1. Mark Sagoff, "On Preserving the Natural Environment," *Yale Law Review* 84 (1974): 205-38; J. Baird Callicott, "Animal Liberation: A Triangular Affair," *Environmental Ethics* 2 (1980): 311-38; and Bryan Norton, "En-

vironmental Ethics and Nonhuman Rights," *Environmental Ethics* 4 (1982): 17-36.

2. Peter Singer, *Animal Liberation* (New York: Random House, 1975), has elaborated a position of this sort.

3. Occasionally, it has been argued that utilitarianism is not just *insufficient* to justify the principles of environmentalism, but is actually mistaken in holding that pain is intrinsically bad. Callicott writes: "I herewith declare in all soberness that I see nothing wrong with pain. It is a marvelous method, honed by the evolutionary process, of conveying important organic information. I think it was the late Alan Watts who somewhere remarks that upon being asked if he did not think there was too much pain in the world replied, 'No, I think there's just enough' " ("A Triangular Affair," p. 333). Setting to one side the remark attributed to Watts, I should point out that pain can be intrinsically bad and still have some good consequences. The point of calling pain intrinsically bad is to say that one essential aspect of experiencing it is negative.

4. See Sagoff, "Natural Environment"; Callicott, "A Triangular Affair"; and Norton, "Ethics and Nonhuman Rights."

5. Callicott, "A Triangular Affair."

6. A parallel with a quite different moral problem will perhaps make it clearer how the environmentalist's holism conflicts with some fundamental ethical ideas. When we consider the rights of individuals to receive compensation for harm, we generally expect that the individuals compensated must be one and the same as the individuals harmed. This expectation runs counter to the way an affirmative action program might be set up, if individuals were to receive compensation simply for being members of groups that have suffered certain kinds of discrimination, whether or not they themselves were victims of discrimination. I do not raise this example to suggest that a holistic conception according to which groups have entitlements is beyond consideration. Rather, my point is to exhibit a case in which a rather common ethical idea is individualistic rather than holistic.

7. Peter Steinhart, "The Advance of the Ethic," *Audubon* 82 (January 1980): 126-27.

8. David Ehrenfeld, "The Conservation of Non-Resources," *American Scientist* 64 (1976): 648-56. For a theoretical discussion see Robert M. May, *Stability and Complexity in Model Ecosystems* (Princeton: Princeton University Press, 1973).

9. See Thomas E. Lovejoy, this volume.

10. See Bryan G. Norton, this volume.

11. Callicott, "A Triangular Affair," p. 330.

12. Elliott Sober, "Evolution, Population Thinking, and Essentialism," *Philosophy of Science* 47 (1980): 350-83; and John McCloskey, "Ecological Ethics and Its Justification: A Critical Appraisal," in *Environmental Philosophy, Monograph Series 2*, edited by D. S. Mannison, M. A. McRobbie, and

R. Routley (Philosophy Department, Australian National University, 1980), pp. 65-87.

13. Callicott, "A Triangular Affair," pp. 333-34 (my emphasis).

14. Paul Shepard, "Animal Rights and Human Rites," *North American Review* (Winter 1974): 35-41.

15. Charles Darwin, *The Autobiography of Charles Darwin* (London: Collins, 1876, 1958), p. 90.

16. The idea that the natural world is perfect, besides being suspect as an ethical principle, is also controversial as biology. In spite of Callicott's confidence that the amount of pain found in nature is biologically optimal (see note 3), this adaptationist outlook is now much debated. See, for example, Richard Lewontin and Stephen Jay Gould, "The Spandrels of San Marco and the Panglossian Paradigm: A Critique of the Adaptationist Programme," *Proceedings of the Royal Society of London* 205 (1979): 581-98; and John Maynard Smith, "Optimization Theory in Evolution," *Annual Review of Ecology and Systematics* 9 (1978): 31-56. Both are reprinted in Sober, ed., *Conceptual Issues in Evolutionary Biology* (Cambridge: MIT Press, 1984).

17. Callicott, "A Triangular Affair," p. 330.

18. Ibid.

19. Sober, "Evolution, Population Thinking, and Essentialism," pp. 360-65.

20. G.E.R. Lloyd, *Aristotle: The Growth and Structure of His Thought* (Cambridge, England: Cambridge University Press, 1968), p. 162.

21. Aristotle, *De Anima*, 415a26.

22. See Sober, "Evolution, Population Thinking, and Essentialism," pp. 360-65. See also Elliott Sober, *The Nature of Selection* (Cambridge: MIT Press, 1984) for further discussion.

23. Ernst Mayr, "Typological versus Population Thinking," in *Evolution and Diversity of Life*, edited by Ernst Mayr (Cambridge, Mass.: Harvard University Press, 1976), pp. 26-29, reprinted in Sober, ed., *Conceptual Issues in Evolutionary Biology*; Richard Lewontin, "Biological Determinism as a Social Weapon," in *Ann Arbor Science for the People Editorial Collection: Biology as a Social Weapon* (Minneapolis, Minn.: Burgess Publishing Company, 1977), pp. 6-20; Sober, "Evolution, Population Thinking, and Essentialism," pp. 372-79; and Elliott Sober, "Darwin's Evolutionary Concepts: A Philosophical Perspective," in *The Darwinian Heritage*, edited by David Kohn (Princeton: Princeton University Press, 1986).

24. Lewontin, "Biological Determinism," p. 10.

25. Christopher Stone, *Should Trees Have Standing?* (Los Altos, Calif.: William Kaufmann, 1972), p. 24.

26. Sagoff, "Natural Environment," pp. 220-24.

27. The example is Sagoff's, ibid.

28. I argue this view in more detail in "Evolution, Population Thinking, and Essentialism," pp. 360-79.

29. Ehrenfeld, "The Conservation of Non-Resources," pp. 651-52.

30. George C. Williams, *Adaptation and Natural Selection* (Princeton: Princeton University Press, 1966); and Sober, *The Nature of Selection.*

31. Aldo Leopold, *A Sand County Almanac* (New York: Oxford University Press, 1949), pp. 224-25.

32. Callicott, "A Triangular Affair," p. 326 (emphasis mine).

33. Edward Abbey, *Desert Solitaire* (New York: Ballantine Books, 1968), p. 20.

34. Garrett Hardin, "The Economics of Wilderness," *Natural History* 78 (1969): 176.

35. Callicott, "A Triangular Affair," p. 323.

36. Ibid., p. 335.

37. John Rawls, *A Theory of Justice* (Cambridge, Mass.: Harvard University Press, 1971).

38. Richard Levins and Richard Lewontin, "Dialectic and Reductionism in Ecology," *Synthèse* 43 (1980): 47-78.

39. Here we would have an inversion, not just a rejection, of a familiar Marxian doctrine—the labor theory of value.

40. Richard Routley and Val Routley, "Human Chauvinism and Environmental Ethics," in *Environmental Philosophy*, p. 154.

41. Mark Sagoff, "On Restoring and Reproducing Art," *Journal of Philosophy* 75 (1978): 453-70.

42. Donald H. Regan, this volume.

43. Routley and Routley, "Human Chauvinism," pp. 121-22.

44. Ibid.

45. I am grateful to Donald Crawford, Jon Moline, Bryan Norton, Robert Stauffer, and Daniel Wikler for useful discussion. I also wish to thank the National Science Foundation and the Graduate School of the University of Wisconsin-Madison for financial support.

8

Duties of Preservation

DONALD H. REGAN

I. The Problem

The central philosophical problem concerning our duties with regard to nature is this: We are strongly inclined to think we have certain duties which are not fully accounted for by instrumental arguments. We are also strongly inclined to hold a view about value that seems to make it impossible to account for these duties by any noninstrumental arguments. Hence our perplexity.

It seems that we have duties to respect living creatures; to avoid causing the extinction of species; even to preserve complex parts of the environment such as a tropical rain forest or the Grand Canyon. If we ask how we can account for these duties (which for convenience I shall refer to collectively as duties of preservation), one possibility is to list the benefits that accrue to us from acknowledging them. There are powerful instrumental arguments for preserving tropical rain forests, which are central to the whole planetary ecology. There are colorable arguments for the general preservation of species, since currently obscure species can turn out to yield new drugs or new genetic material for agricultural exploitation. There are more sophisticated arguments of potentially unlimited application based on "aesthetic utility"—we should preserve particular natural features or particular species because they embody qualities we value or because we find them pleasing. Instrumental arguments like these are numerous, varied, and significant, but they do not satisfy us entirely. They entail the consequence that if a species is not ecologically essential, and if it will produce no wonder drug or wonder crop, then whether we have a duty to preserve it depends on whether we like it. The content of our duties of preservation will depend heavily on accidental features of human preferences. This seems wrong. Analogously, even in the case of the tropical rain forest, whose present "economic" utility is not in doubt, it seems wrong to claim we should preserve it merely because it is essential to our carrying on our lives and pursuing our goals and satisfactions, whatever they may be. The day may come when we can abandon this planet and dispense with the rain forest. We do not believe, however, that on that day it will become a matter of moral indifference whether we destroy the forest. Our duties of preservation demand a more solid foundation than our own changeable needs and fancies.

On the other hand, suppose we try to account for duties of preservation without such instrumental arguments. It seems we must claim the natural world is valuable in itself. Unfortunately, that claim conflicts with another common intuition, that the existence of value depends somehow on the existence of fairly sophisticated consciousness. If we imagine a universe in which there is no life form higher than the nematode, and if we ask whether it is better for that universe to exist than for there to be nothing at all, many of us would be inclined to say that it is not better—between the existence and the nonexistence of such a universe, there is nothing to choose.[1] I do not say at this point that the existence of value depends on the existence of human beings. If value depends somehow on consciousness, then it remains to be determined what level and type of consciousness is required and whether we are the only creatures that possess it. But these questions are not crucial here. Our duties of preservation seem to extend to natural entities that plainly do not have conscious lives of their own of any sophistication—the nematode, the lousewort, the Grand Canyon. It is hard to see how we can explain the duties of preservation in a noninstrumental fashion without claiming these entities are valuable in themselves and thereby contradicting the intuition that value and consciousness go together.[2]

II. On Not Avoiding the Problem

The problem, then, is to give a noninstrumental account of our duties of preservation without denying the connection between value and consciousness. We could, of course, avoid the problem without solving it. We could deny the supposed duties of preservation, or we could grant intrinsic value to entities which lack consciousness. I propose to solve the problem, not avoid it. However, my solution may seem so implausible as to suggest that one of the courses of avoidance must be preferable. I therefore want to say something against each of the courses of avoidance. It will make for the smoothest exposition (even though it will upset the symmetry of the presentation) if I say a bit more now about the connection between value and consciousness and put off until much later my reasons for not simply denying duties of preservation.

With regard to value and consciousness, I propose to look briefly at three standard attempts to explain duties of preservation and to show that while none of them is satisfactory, each of them achieves what plausibility it has by arguing in a way which tends to impute consciousness where it does not exist. The lesson is that hardly anyone is willing to face the claim that value depends on consciousness and deny it outright.

(1) It is often said that a variety of nonhuman entities—animals, plants, rivers, the "biotic community"—have rights. That such claims should be made was inevitable; "rights" have become the philosophers' stone of modern moral discourse. Unfortunately, we have no satisfactory understanding of what we mean even by claims that humans, the paradigmatic right-holders, have rights. It is this basic lack of clear understanding that encourages claims about nonhuman rights.

It often seems that claims about Jones's rights are merely disguised claims about other people's duties regarding Jones.[3] If talk about rights is merely disguised talk about duties, then rights cannot be the basis for duties. This is as true of duties regarding nature as of duties regarding humans. On the other hand, if we try to find a notion of rights that has enough substance to be a genuine basis for duties, we are driven to thinking about rights as protections of some sort for the activities of rational, moral agents. It is clear that few or no nonhuman entities have rights of this sort.

What I have said makes it clear that talk of nonhuman rights serves no useful purpose, except perhaps in rare cases. Such talk survives because our fuzzy-mindedness about the nature of rights allows assertions about rights to seem significant when they are not, when they are mere covers for assertions about duties. The seeming significance depends on the suggestion, easily recognized as false in the case of almost all nonhumans once it is made explicit, that the supposed right-holder is a rational, moral agent.

(2) It is often said that we have duties to every living creature because every living creature has "a good of its own."[4] This argument does nothing to explain any duties we might have to inanimate nature. Beyond that, the argument gets a good deal of its apparent force by inviting confusion.

Consider an oak tree. If someone who says that an oak tree has a good of its own is pressed to justify that claim, he will usually respond by pointing out that an oak tree is a complicated, self-regulating, self-sustaining system. Now, if that is what it means to say that an oak tree has a good of its own, I agree that an oak tree does have a good of its own. But why does it follow that I have any moral duty regarding the oak tree? The claim must be that, other things being equal, I should not interfere with complicated, self-regulating, self-sustaining systems. As it happens, I think this claim has some plausibility, for reasons that will become clear further on. Even so, this claim is no place to end the argument for duties regarding nature. It is at most a suggestive place to begin.

The real problem with saying that an oak tree has "a good of its own" is that it leads us almost inevitably to think of the oak tree as engaged in conscious valuation and in conscious striving after the "good" it has chosen or acknowledged. Hardly anyone would make such claims about the oak tree outright. But there are considerable illicit benefits (from the point of view of one trying to justify duties regarding the oak tree) from having these ideas in the air. Thinking of the oak tree as a valuer helps us over our worry about how there can be value in nature considered totally apart from human beings (or rational consciousness). We can concede that the existence of value requires a valuer, but we can let the oak tree fill that role. The idea is that each living thing creates the value in its own life by valuing it. Also, we readily think we should not try to impose our views of the good on other people; each person has a right to "a good of his own" which others should respect. If so, and if an oak tree has a good of its own, then it seems we should respect that too. These benefits are, as I said, illicit. The oak does not engage in conscious valuation. If we think it important that an oak tree is a complicated, self-regulating, self-sustaining system, let us say that about the oak tree and then argue about why it matters. Let us avoid specious shortcuts. (What I have said in this section applies *mutatis mutandis* to claims that every living creature has "interests," or that every living creature is a center of a respect-demanding "will.")

(3) It is often said that we are members of a "global community" which includes, in addition to ourselves, all other living creatures from chimpanzees to blue-green algae, and it is said that we have duties to the other members of this community just as each of us has duties to other members of the human community.[5] However, "community"—in the only sense in which it can possibly have any moral significance—requires at least the potential for shared beliefs and values. The universe of living creatures simply does not amount to a community in any morally relevant sense.

The "global community" usually turns out to be a metaphor for the fact that all life is part of a complex ecological web. We are dependent on many other species; and almost every other species (perhaps absolutely every other species) is dependent on us in the sense that we could destroy it if we tried. What follows from these facts? Nothing of interest. Our power to affect other species may be a necessary condition for our having duties regarding them; but the mere fact of our having this power does not tell us what duties we have, if any, in the use of it. Similarly, our dependence on other species means that if we do too much damage to other living things we are likely to end up hurting ourselves as well; but any argument from this premise will make our duties regarding nature instrumental, which is just what the argument from global community was meant to avoid.

III. The Solution

It is time now to introduce my own solution to the problem of justifying noninstrumental duties of preservation without denying the connection between value and consciousness. I shall borrow an idea from G. E. Moore. The idea is best introduced by considering a specific example. There is a natural "object" which is the Grand Canyon. Many people know of the existence of the Grand Canyon and have some specific knowledge of it. Consider a person Jones, who has some knowledge of the Grand Canyon and who also takes pleasure in that knowledge. Moore's idea is that we should regard as intrinsically valuable the *complex* consisting of the Grand Canyon, plus Jones's knowledge of the Grand Canyon, plus Jones's pleasure in her knowledge of it.

It must be emphasized that what is valuable here comprises three things—the Grand Canyon, Jones's knowledge, and Jones's pleasure in her knowledge of the Canyon—when they occur together. The valuable entity is what Moore called an "organic unity." The claim that this unity is valuable does not, at least on the face of it, entail that any individual element is valuable on its own. In particular, it does not entail either that the Canyon is valuable on its own or that Jones's pleasure is valuable on its own.[6]

The phrase "Jones's pleasure in her knowledge of the Canyon," which I have used to describe one element of the valuable complex, may be misleading. It suggests that Jones's pleasure is in her knowledge alone, and not in the Canyon. It suggests that Jones is glad to know about the Canyon, given that it is there, because she likes knowing about what is there, but that she is otherwise indifferent to the existence of the Canyon. That is not the state of mind I want Jones to have. I want Jones to take pleasure in the existence of the Canyon itself. Why, then, do I not use the phrase "Jones's pleasure in the existence of the Canyon"? Because that also may be misleading. This latest phrase suggests that, so long as Jones knows there is a Grand Canyon, and is glad of that, it does not matter whether Jones knows or cares to know anything more about the Canyon. On my view, however, Jones ought to want to know as much about the Canyon as possible; and she ought to take greater pleasure in knowing it more completely; and the more she knows the more valuable is the complex under discussion. Perhaps, then, we should speak of "Jones's pleasure in the existence of the Canyon and in her knowledge of it." This is an improvement, but it still could be taken to refer to two separable pleasures in two separate things, the Canyon, and Jones's knowledge of it. The misleading suggestions of the previous inadequate phrasings are not suppressed entirely. Perhaps the best description of Jones's state of mind would be "Jones's pleasure in the Canyon as known to her." The only trouble with this is that it sounds rather artificial. We would hardly know what it meant if it did not come at the end of a paragraph of hair-splitting. Hereafter I shall use all the phrases I have discussed indiscriminately. I shall always mean by them what I would mean by "Jones's pleasure in the Canyon as known to her"—that is, a pleasure in the Canyon and in Jones's knowledge of it as an inseparable whole, and a pleasure which would be magnified by greater knowledge.

So far I have described one thing, a complex involving the Grand Canyon and Jones, which I have claimed is intrinsically valuable. Shortly I shall give a recipe for constructing enormous numbers of valuable complexes along the same lines. First, however, let me say a bit about what it means to call something "intrinsically valuable."

To say something is intrinsically valuable is to say that it is good in itself, abstracting from its causal antecedents or consequences, and abstracting from its relationships to other things. It is good just for what it is. A universe which contains it is the better for containing it (still abstracting from causes, consequences, and external relationships).

If something is intrinsically valuable, then any moral agent has a moral reason to try to bring it into existence or to preserve it if it exists already. Of course, to say a moral agent "has a moral reason" to promote the existence of something is not to specify the strength of that reason; and it is not to say that the agent has a *conclusive* moral reason. Two things may both be intrinsically valuable, and yet one may be more valuable than the other. If we must choose between the existence of two valuable things, of different value, we have "a reason" to promote the existence of each, but we have the stronger reason, and thus the conclusive reason, other things being equal, to prefer the existence of the more valuable. Similarly, something may be intrinsically valuable and also be instrumentally valuable or instrumentally disvaluable. (Of course, this instrumental value or disvalue must ultimately depend on the intrinsic value or disvalue of something else.) We may have a conclusive reason, all things considered, to avoid creating or even to destroy some complex of great intrinsic value, if its existence has sufficiently bad consequences. All of this is elementary; but people sometimes assume that if a thing has intrinsic value it must have "absolute value" (either infinite value or value that for some other reason cannot be weighed against competing values), or else they assume that if a thing has intrinsic value it cannot have instrumental value or disvalue as well. These are misconceptions.

How does one go about showing that something has intrinsic value? One does it in exactly the same way one demonstrates any other fundamental moral proposition, however that is. I do not claim to know how that is. I do claim that the widespread belief that propositions about the good are harder to demonstrate than propositions about rights and duties is the merest prejudice. For the present what we can do is to expose and avoid outright confusions; to consult our moral intuitions; and to see which intuitions (at all levels of particularity or generality) figure in the most coherent moral theory we can come up with.

Very well, what other complexes, constructed on the model of the Grand Canyon–Jones complex, are intrinsically valuable? Clearly, if we replace Jones by any other human being, the resulting complex will still be valuable if the original one is. Indeed, if we replace Jones by any other conscious agent capable of having the same sort of knowledge of the Canyon and taking the same sort of pleasure in it as Jones does, the resulting complex will still be valuable. The interesting question is what we can put in the place of the Grand Canyon. I do *not* suggest that we can put in the place of the Grand Canyon anything at all which we can refer to with a definite description. I would not, for example, claim that the complex consisting of Smith's cruelty, plus Jones's knowledge of Smith's cruelty, plus Jones's pleasure in her knowledge of Smith's cruelty, is intrinsically valuable.

On the other hand, I suggest that we can put in the place of the Grand Canyon any "natural object." "Object" here is to be construed broadly, as including anything we can refer to with a definite description. Thus, natural objects include individual creatures, species, individual ecosystems, types of ecosystems, geological formations, and natural processes (everything from "the way prokaryotes exchange energy with their environment" to "the way continents have been created and destroyed by plate tectonics"). Pions and muons are natural objects, and so is the process of heavy-element formation in stars. In saying that all these things are natural objects, I do not mean to be making any controversial metaphysical claim. I am saying only that every one of these things can be part of an intrinsically valuable complex if it is conjoined with some human being's knowledge of it and that human being's pleasure in her knowledge. (I shall speak for now as if the only relevant knowers are human beings.)

With Moore's idea, as I have elaborated it, in hand, we have the materials for an argument for the preservation of any natural object. Any natural object can be part of a valuable complex if we know the object and take pleasure in our knowledge. As to any such complex, we have a reason to promote its existence. If the existence of the object is essential to the existence of the complex, then we have a reason to promote the existence of the object. So we have a reason to preserve any natural object that currently exists.

Now, consider some particular natural object. Someone might say: "Unless we know the object and take pleasure in our knowledge, the object's mere existence is not valuable. Therefore our having a reason to preserve the object depends on our knowing it and taking pleasure in it. We have no reason to preserve an object we do not know, or that we take no pleasure in. So what we should preserve depends on what we like, and we are back where we started." This is not correct. It is true that if our ignorance of the object, or our lack of pleasure in it, were an immutable feature of the universe, then we would have no reason to preserve the object in question. But we are capable of learning about the object; and I believe we are capable of learning to take pleasure in knowledge of any natural object. If that is so, then we have reason to preserve any natural object, just as I said. With regard to objects we already know about and take pleasure in, it is enough to preserve them. With regard to objects we do not know or take pleasure in, we have reason to preserve them, and to study them, and to learn to enjoy our knowledge of them. (Even as to objects we already know something about and enjoy knowing about, there is probably always more we might know. If, as I assume, fuller knowledge makes a more valuable complex, then we have reason to study these objects as well.)

Observe what my argument, if we can accept it, accomplishes. It gives us a noninstrumental argument for the preservation of natural objects, without denying the intuition that the existence of value depends somehow on consciousness. Both of these points bear spelling out.

IV. On Noninstrumentality

The main thing I mean when I say the present argument is "noninstrumental" is this: On the present argument it is not the case that we have reason to preserve a natural object only if it is economically useful or if we happen to like it. We have reason to preserve every natural object; those we do not currently have knowledge of or take pleasure in, we ought to learn about and learn to take pleasure in. It is true that our having reason to preserve every natural object depends on the fact that we have at least the potential for knowing about and taking pleasure in it. But that is a potential we have and which, on the present argument, we ought to develop. The argument for preservation does not disappear just because we decide to ignore that potential. Nor is it necessarily overridden if it turns out that we could satisfy more of our preferences overall by ignoring that potential. Satisfying our preferences is not all that morality is about. The central objection to traditional instrumental arguments for preservation is that they make the case for preservation depend on accidental or arbitrary facts about our needs and preferences. The present argument depends on only one deep fact about us, which, while it could be otherwise, seems too fundamental to be regarded as accidental or arbitrary. The fact that we are able to take pleasure in the knowledge of any natural object is not at all similar to the fact that most of us like butterflies and do not like slugs.

The present argument is also noninstrumental in the sense that it does not make preservation a means to maximizing "the good for man." The good to be maximized is one in which human beings figure essentially (assuming we are the only sufficiently sophisticated consciousnesses in the universe). The good complexes I have described cannot exist without certain facts being true of certain humans. But these complexes are not good because they are "good for" the humans involved.

It might seem that my argument is instrumental in the sense that the existence of any natural object is valuable only because it is necessary to something else, namely, the existence of various complexes of the sort I have described. This suggestion rests on a confusion about the nature of instrumental value. It is true that the existence of a natural object is valuable only as part of one of my complexes. But that is not to say that its value is merely instrumental. Its value does not lie in its consequences; it is not valuable because it causes something else. The complex which is valuable is different from the object alone; but the object is an essential part of it. The relationship between the object and what is valuable in itself is a logical relationship, and not a merely causal one. (I should perhaps point out that the object may have instrumental value, even with respect to a complex in which it figures. The object must exist before it can be known. Its existence at one time may have as a causal consequence its existing and being known at a later time. So its existence at the earlier time may be instrumental to the existence of the valuable complex at the later time. It remains true that the value of the object is not *merely* instrumental, since the object figures essentially in the valuable complex realized at the later time.)

V. On Value and Consciousness

I have spelled out the point that my argument for preservation is not instrumental. Let me say a word about the point that it is consistent with our intuition about the relation between value and consciousness. Someone who says that the existence of value depends on consciousness may mean two quite different things. He may mean, "No proposition about value is true except in virtue of being believed by some conscious being." Alternatively, he may mean, "Nothing has positive or negative value unless it somehow involves consciousness." The first claim is metaethical: it is about truth-conditions for moral claims, and it entails subjectivism of some sort. The second claim is normative: it is about what sort of thing in fact has value, and it is consistent with objectivism. The simplest way to see the difference between these claims may be to imagine a universe devoid of consciousness. The proponent of the first claim might well say about that universe that it has "no value" in the sense that no attribution to it of any particular degree of value is true. Because the universe is devoid of consciousness, the question of its value simply cannot arise. (The proponent of the first claim need not say this, but it is something he could say.) The proponent of the second claim, in contrast, would say that the universe without consciousness has a quite definite value, namely zero.

The claim I am presently interested in is the second, normative, claim. It is the second claim that, in the opening paragraphs of this essay, seemed to stand in the way of explaining our duties of preservation by noninstrumental arguments. But it is clear that my Moorean argument for preservation is perfectly consistent with the intuition that nothing has value unless it somehow involves consciousness. The only things that I have asserted are valuable *do* involve consciousness. The consistency of my view with the claim that value requires consciousness is obvious, once we are clear about what version of that claim creates the difficulty.

VI. Objections and Complications

I have said that a natural object such as the Grand Canyon is not valuable in itself, but that it is good for the Canyon to exist and for us to know about it and take pleasure in our knowledge. Someone might object that this position is incoherent. How can it possibly be a good thing for us to know about the Canyon and take pleasure in it unless the Canyon is worth knowing about and taking pleasure in? In short, does not my view, that a certain sort of complex involving the Canyon is good, presuppose that the Canyon is good in itself?[7]

I concede that on my view the Canyon must be, in some sense, "worth" knowing about and taking pleasure in. It does not follow, however, that the Canyon must be valuable in itself. Rather, to say the Canyon is worth knowing about is to say precisely that complexes involving the Canyon and someone's knowledge of it and pleasure in it are good. To say this is to assert something about the Canyon which is not tautological or trivial. As I have already pointed out, there are things (such as "Smith's cruelty") which do not generate good complexes when someone knows about them and takes pleasure in them. So, the claim that the Canyon is worth knowing about is a significant claim even though it does not entail that the Canyon is valuable in itself.[8]

The hypothetical objector may not be satisfied. How do I know that the complex involving the Grand Canyon is good and the analogous complex involving Smith's cruelty is bad? Must not there be something about the Grand Canyon (itself) and Smith's cruelty (itself) that accounts for this difference? What could it be except that the Canyon is good in itself and Smith's cruelty is bad in itself? A complete answer to this question would take us far beyond the present essay. The short answer is this. There is a difference between the Canyon and Smith's cruelty, and it is a difference in intrinsic value. But the difference is that while Smith's cruelty is bad in itself, the Canyon has zero value in itself. (Note that Smith's cruelty can be bad in itself without violating the principle that value requires consciousness, since Smith's cruelty is itself a feature of a certain consciousness.) As to things, like Smith's cruelty, that are good or bad in themselves, it is good to know them and take pleasure in them if they are good, bad to know them and take pleasure in them if they are bad. (It is not the knowledge but the *taking pleasure* that makes the complex bad when the thing is bad in itself.) As to things, like the Canyon, that are neither good nor bad in themselves, knowing them and taking pleasure in them is good. There is an asymmetry in this position. But that, so far as I can now make out, is the way it is.

A different objection to my claim that the Grand Canyon has no value on its own depends on the well-known "last person" argument.[9] Suppose Jones is the last person on earth. She knows there will be no one after her. As she dies, she can push a button which detonates a cataclysmic explosion and destroys the Grand Canyon. Would it be wrong for her to push the button? Many people are inclined to think it would be wrong; and they also think that the wrongness of pushing the button can only be accounted for by granting the Canyon value in itself. After all (the argument goes), once Jones is dead, the universe is empty of consciousness. If consciousness is essential to value, then there is zero value, whether the Canyon is still there or not. If it makes no difference to the value of the universe whether the Canyon exists after Jones dies, she can hardly do wrong by destroying it in the moment of her death.

My own view is that it would be wrong for Jones to destroy the Canyon, but that we can account for this consistently with my theory. There are two different points to be made, both of them important. First, assuming Jones destroys the Canyon, why does she do it? We could describe a variety of possible motivations, which differ subtly, but there must be a common element of sheer wanton destructiveness. Jones wants the Canyon not to exist. But that desire is inconsistent with Jones's knowing of the existence of the Canyon and taking pleasure in it. If Jones has the proper attitude toward the Canyon (that is, the attitude which would figure in a valuable complex involving the Canyon and herself), she will not destroy the Canyon. Conversely, if she does destroy the Canyon, she has the wrong attitude, and that is what is wrong in the situation. It is true that on this analysis her act of destruction is wrong only derivatively. The act is wrong because it manifests a wrong attitude, while the attitude would be wrong whether the act were done or not. Still, it is the act that manifests the attitude to us, so it is natural that we should locate the wrongness in the act, at least on first consideration. (Someone might ask us to imagine that Jones does take pleasure in the existence of the Canyon, so long as she exists, but that she is saddened by the thought of the Canyon surviving her. She wants to affirm her intimacy with it by "taking it with her." I do not find this psychologically plausible. If it were plausible, then perhaps we should say that Jones was extemely peculiar, but in such a way as to make her apparently wrong act arguably innocent.)

My second point about the "last person" argument avoids any psychological claims, but raises other complications. The argument that it does not matter whether the Canyon exists after Jones's death assumes that the Canyon's existence can only be valuable as part of a complex involving contemporaneous knowledge by some consciousness. Is it not possible, however, that the value of the universe could be increased by the Canyon's existence after Jones's death, combined with Jones's knowledge (before her death) that the Canyon will continue to exist and her pleasure in that knowledge? When I first described the sort of complex we have been discussing, I ignored this matter of timing, and I must admit that it raises many questions I find perplexing. But we plainly cannot insist on contemporaneous consciousness in every case. Consider human knowledge of dinosaurs. No dinosaur was ever known by a contemporary human consciousness, but it is good that dinosaurs should have existed and that humans now should know about them and take pleasure in that knowledge. If there is good to be achieved by knowledge of what came before the first human, why not also by knowledge of what comes after the last one? But if there is good to be achieved by knowledge of what comes after the last human, then Jones, in destroying the Grand Canyon, destroys an opportunity for good.[10]

Now we have a new problem. I have said that our knowledge of dinosaurs is valuable even though dinosaurs no longer exist. That might seem to affect the nature of our obligation to preserve species that do exist. Imagine that there is some species about which we know everything there is to know. Is there really any reason to preserve that species? We will not learn more about it if it continues to exist. If it ceases to exist, we can remember it. We can continue to know "how it was" and to take pleasure in our knowledge. What difference does it make whether the species is still around?

Although I think it is not a matter of indifference whether the species continues to exist, I should admit that in one respect the case for preservation is *weakened* by our complete knowledge of the species. If we find ourselves in circumstances where we must choose between two similar species—where both are endangered but only one can be saved—we have reason to save the species about which we know *less* in order that, after study, we may eventually end up with the greatest aggregate knowledge of the two species.

Going back to the case where there is just one species which we know completely, why is it desirable that that species continue to exist? One point, of course, is that its continued existence may be important instrumentally to our continuing knowledge of its past existence and pleasure in that knowledge. This argument disappears, however, if we assume that our memory simply will not fade. Another tempting suggestion is this: The species cannot be completely known if it ceases to exist, since if it continues to exist it will evolve and will have a future development that we cannot predict. Therefore the question why it is good for a completely known species to continue to exist cannot properly arise. This may be the right answer, but it raises complications. What are the criteria for identity of a species over time? If the species evolves, is it the same species? If it gives rise to a different species, then the consequence of continuing its existence past the present is not that it will continue to exist and be known by us, but that something new will come into existence which we can know. Our knowing that new thing would, on my view, be a source of value; but if the present species ceases to exist for some reason other than self-annihilation by evolution, then presumably *different* new things will come into existence for us to know. It is not immediately obvious which possible set of new things would, if known, generate complexes of greater intrinsic value.

For the present, I am inclined to rest simply on an intuition that it is better, other things being equal, for something to exist and be known than for it not to exist and (though nonexistent) be known. Many philosophers have had the intuition, in one form or another, that existence is good. I have rejected the claim that the mere existence of entities without consciousness is a good. But that rejection is perfectly consistent with the suggestion, which I would expect to be widely accepted and which reflects the intuition that existence is better than nonexistence, that existing-and-being-known is better than non-existing-and-(nonetheless)-being-known.

We should also remember that the question "What difference does it make whether the perfectly known species continues to exist?" will arise in some context. Insofar as we are called upon to make a choice, our choice can be expected to manifest our attitude toward the species, just as Jones's choice about the Grand Canyon manifested her attitude. If we now take pleasure in our knowledge of the species, we will not be able to regard it as a matter of indifference whether the species continues to exist. We may decline to save it, or we may even eliminate it, for adequate cause. But we will not let it go or eliminate it for no reason at all.

VII. SOME APPLICATIONS

So far I have been engaged in general exposition of my theory. It may clarify how the theory works if we look at some examples.

(1) Consider a case where we must choose between protecting two individuals of some common species and protecting the last two (mating) individuals of a similar but rare species. Most of us would think it better to protect the rare individuals. Theories based on the "rights" or "interests" of individual living creatures cannot account for this intuition.[11] My theory, however, accounts for it very simply. Whichever pair of individuals we protect, we save two individuals (whose existence we know of and take pleasure in). That is good. But if we protect the rare individuals, and only if we protect them, we also save a species which would otherwise disappear. Because the species is a separate "natural object," saving the species is a separate good. What I have just said may seem too simple to be an argument. I have hardly done more than restate the obvious basis for our intuition that we should save the rare individuals. The point is that my theory, on which every natural object is potentially valuable, and valuable in essentially the same way though not necessarily to the same degree, provides a natural home for our intuition. Competing theories do not.

(2) Consider a case where we must decide whether or not to prevent the extinction of a species which seems to be disappearing in the natural

course of events and not because of human intervention. Even people who feel strongly that we should not cause extinctions are likely to feel that allowing an extinction to occur (not preventing it when prevention is in our power) is quite a different thing. This obviously cannot be explained by a theory which assigns value only to species as such. On my view, the difference between causing an extinction and allowing an extinction is explained by the fact that natural processes are potentially valuable natural objects just as species are. Where we cause an extinction, we are destroying both the species and its natural "trajectory." If we prevent an extinction, we are saving a species (which is good), but we are interfering with a natural process. It does not follow from this that we should always allow extinctions. There is a choice to be made. But when the issue is whether to allow an extinction, as opposed to whether to cause an extinction, there are considerations on both sides.

(3) A similar question is whether we should interfere in a particular case to save an animal about to be eaten by a predator. On any theory which confines its attention to individuals, it is hard to see why we may not choose either way. The same is true even on a theory that values species, if the choice we are considering arises in only a small number of cases. However, many people would think that, barring special circumstances, we probably should not intervene. This again is easily explained if we grant value (by which I always mean potential value as part of an appropriate complex) to the natural processes of predator-prey interaction.

(4) A rather different question is whether we should preserve things which are dangerous, such as the smallpox virus or rats, or things which are generally thought ugly, like banana slugs. My theory says that we have reason to preserve the smallpox virus and that the reason is of the same type as the reason for preserving, say, butterflies. But if the consequential costs of preserving the smallpox virus are too great, then on balance we should extinguish it. Calculating the consequences and striking the balance may be very complicated. But the fact that the smallpox virus is dangerous to people does not mean it is not worth knowing about or that we may not take pleasure in knowing about it. No more does the "sliminess" of the slug make it unworthy. "*All* God's creatures" are worth our attention.

(5) Sometimes we have to destroy something, like a laboratory animal or the natural operation of some natural process over a particular stretch of time, in order to increase our knowledge of the thing we destroy or of similar things. On my view, this destruction may be justified. The value of our knowledge of and pleasure in anything is increased if the knowl-

edge increases. In some cases, then, the destruction, which should always be cause for regret, will produce adequate compensating benefits.

(6) There has been discussion of whether we should focus our efforts at preservation on rare species or on rare ecosystems.[12] There is even the suggestion that protecting species and protecting ecosystems will come to the same thing in practice. Although I agree with much of what is said about this, I would point out that there is a difference in principle between protecting species and protecting ecosystems. We can easily imagine an ecosystem which contains only common species (and only species common outside that ecosystem) but which is nonetheless special as an ecosystem because the species interact there in unique ways. Similarly, we can imagine a species which is rare, and which inhabits only one ecosystem, but which does not make the ecosystem as such specially interesting because a similar ecological role is played by a similar but more common species in many similar ecosystems. Given that there is a difference between protecting rare species and rare ecosystems, my theory says that we should protect both. I think that comports with most people's intuitions.

(7) Finally, what is the value of human-created species? This is an unusual question, but a very interesting one. It seems to me that human-created species must be regarded as human artifacts. Many human artifacts are valuable in the same general way as natural objects; but they are not valuable in exactly the same way. To explain why not, I must add to my theory a point that has not been brought out previously. I have not attempted to characterize in full detail the emotion I have referred to as our pleasure in the knowledge of various natural objects. One aspect of that emotion ought to be pleasure in the knowledge of them as natural, as things not of our making. Our relation to the natural world is complex. We are rooted in it. We evolved out of it. But also, in some sense, we have evolved *out* of it. Unless we are in some deep way different from earthquakes, oak trees, and lemurs, moral philosophy, and indeed all philosophy and all science, is an illusion. I do not claim to understand just how we are different from earthquakes, oak trees, and lemurs; but if there is such a gulf as I claim between the human world and the natural world, that is itself a fact (or an enormous system of facts) worth learning about and taking pleasure in. It follows, I think, that we should value what is natural in part for being natural. And so, of course, we should value what is human in part for being human. Everything made by humans (or at least every material thing made by humans) has a natural component; and the proportions and mode of mixing of the natural and the "human" may vary greatly. I do not mean to suggest that a "wild" garden and a freeway interchange are to be

valued in exactly the same way because they both show the human hand. But the distinction between what is and what is not our creation would be crucial to a complete explication (which I cannot here provide) of how we should enjoy the world around us.[13]

VIII. PRIORITIES AMONG SPECIES

I turn now from specific cases to some general remarks on the problem of priorities among species. These remarks will also tend to fill out my theory. The question is this: If we can preserve only a limited number of endangered species, and if we have some choice about which to preserve, how should we decide? Specifically, what makes one species intrinsically more valuable than another? The second question, about intrinsic value, is narrower than the first question, about how we should decide. I shall focus on the second question, but I should point out what I am ignoring, which is questions about the instrumental value and disvalue of species.[14] Species often have substantial instrumental value or disvalue. It is likely that in many cases where we must decide what to preserve, the instrumental considerations completely swamp considerations relating to the intrinsic value of the species themselves. Of course, in principle, we cannot know what the instrumental considerations are without having a complete theory of intrinsic value. But a complete theory would recognize many forms of intrinsic value not discussed in this essay. So, in discussing only the intrinsic value of species, I am dealing with only a part, perhaps a very small part, of the problem.[15]

It seems that there are two clear general principles. First, more complex life forms are more valuable than less complex ones, other things being equal. A ferret is more valuable than a centipede, which is more valuable than an amoeba. Even an amoeba is enormously complex. Still, there are degrees of complexity; and that remains true even if I am wrong about my specific rankings. Given that there are degrees of complexity, the more complex object is more valuable just because it provides the opportunity for more knowledge. There is more to be known about a complex object than about a simple one. (To be sure, the possibility of the thing being "known" at all depends on its having a certain stable internal organization. Complexity is not the same as chaos, and there is a sense in which organization is simplicity. But these facts do not seem to upset our intuitions about the value of complexity and the reason it is valuable.)

In the preceding paragraph I compared "a ferret" with "a centipede," and so on. The original question concerned the value of a species of ferret or a species of centipede. Does the value ranking remain the same whether we talk about individuals or species? In general, yes. There may be some exceptions, however. Social insects, for example, exhibit complexities of behavior as species that would be unknown to us and possibly undiscoverable if we knew these insects only as individuals. So in this case the species is relatively more valuable than we might think after considering only the (apparent) complexity of individuals. Similarly, certain species may play special roles in complex and unusual ecosystems, which makes a complete knowledge of them and their systematic role more valuable than it would otherwise seem.

The second clear general principle is that, other things being equal, a species is made more valuable by belonging to a sparsely populated genus or family or order. Taxonomically isolated species are likely to represent unusual modes of adaptation. Their existence increases the diversity of nature. That means both that there is more to know about what exists, and that a complete knowledge of what exists entails a greater knowledge of nature's possibilities.

Needless to say, the values of complexity and diversity may conflict. We might be compelled to choose between a complex species in a richly populated genus and a simpler species which has no congeners at all.

A question that I shall raise without answering is whether a species acquires special value by having a role in some human culture. Animals figure in our cultural practices in a variety of ways. The bald eagle is the national symbol of the United States. Certain Eskimo tribes depend heavily on the bowhead whale. English-speaking children (and many adults who were once English-speaking children) may have special feelings about rabbits and deer because they were brought up with Peter Rabbit and Bambi. And so on. Clearly any of these facts may invest a species with special instrumental importance. The question is, do any of these facts invest the species with special intrinsic importance? Or rather, since we deny intrinsic importance to the species itself, do any of the cultural complexes we have indicated have an intrinsic value that is not exhausted by the value of the "natural object plus knowledge plus pleasure in the knowledge" and in which the species figures essentially? It should be apparent why, having raised this question, I cannot answer it here. An answer would presuppose a complete theory of the intrinsic value of cultural manifestations. We would also need to say more than we have said about the special value (if any) of shared knowledge, as opposed to knowledge belonging (or considered as belonging) to individual knowers only. I suspect that in the end we would conclude that the intrinsic value of a species is increased by its playing a special role in human culture. But for now that must remain a conjecture.

IX. THE IMPORTANCE OF BEING HUMAN

For much of this paper I have written as if the valuable complexes under discussion must involve, in addition to some natural object, a human being who knows the object and takes pleasure in it. Is this concentration on human knowers justified? As I have already explained, it is not my intention to limit the relevant range of knowers by stipulation. In principle, other knowing subjects could fill the role of the human being in the complexes I regard as valuable. It just seems to me that in the universe as we know it, there are no other subjects who can fill the same role.

Why not? Obviously, there is both a normative and an empirical premise. The normative premise specifies just what kind of knowledge and pleasure—just what sort of consciousness—is necessary to the goodness of one of my complexes. The empirical premise is that there are no other subjects in the universe who possess the relevant sort of consciousness.

I have said the subject whose participation makes a valuable complex must know some natural object and take pleasure in it. We cannot deny that certain creatures besides ourselves have "knowledge" of some sort and experience "pleasure" of some sort. But knowledge and pleasure come in many varieties. Although I cannot define precisely the features of consciousness that go to make a valuable complex, I can make it clear that the consciousness must be relatively sophisticated.

First, the knowing subject must be self-conscious. He (let me call the indefinite and possibly nonhuman subject "he") must be aware of himself as something separate from the object of his knowledge. Why is this important? Because the subject must take pleasure in the object and in his knowledge of the object. The subject who does not distinguish between himself and the object may feel pleasure as a result of his awareness of the object, but it seems that he can hardly take pleasure in the object (as such). Even more clearly, unless he is aware of himself as a knowing subject, he is not aware that he has knowledge, and so he cannot take pleasure in his knowledge.

A related point is that the subject must be able to contemplate the possibility of the nonexistence of the things he takes pleasure in. To take pleasure *in* something, as opposed to merely getting pleasure from something, requires that one consciously prefer the existence of what one takes pleasure in to its nonexistence, and therefore requires that one be able to contemplate its nonexistence.

If the full range of possible goods from knowledge of and pleasure in natural objects is to be achieved, the knowing subject must be capable of wide-ranging intellectual apprehension. For example, if the full range of possible goods is to be achieved, the subject must be able to appreciate

213

modern physics. This obviously requires an intellectual capacity considerably in excess of day-to-day needs. It also requires the ability to engage in complicated symbolic activity.

Finally, I am inclined to think that the knowing subject must at least be able to raise the question of what is intrinsically good. Whether this is a further required feature of the relevant consciousness or whether ability to raise value questions is entailed by the other features I have mentioned and briefly argued for (self-consciousness, ability to contemplate nonexistence, wide-ranging intellectual apprehension and symbolic capacity) is itself an interesting problem. Whatever the answer to that problem, however, the reference to ability to raise value questions suggests a useful way of characterizing the required features as a group. Even if the first four features are not sufficient to guarantee the ability to raise value questions, they seem necessary to that ability. In a sense, then, the whole package adds up to the ability to raise value questions. This seems eminently reasonable. Remember that what got us started on this search was a feeling that the existence of value somehow depends on consciousness. If that is so—if we are trying to define the sort of consciousness which is essential to the existence of value—then it seems fitting that the relevant consciousness should turn out to be one that can appreciate value questions.

As to the empirical question whether humans are the only subjects in the universe who have the relevant sort of consciousness, I do not claim to know for certain. No one can afford to be too confident today that he knows all there is to know about the intellectual life of chimpanzees or porpoises. I would claim that, on our planet, human beings are currently far beyond any other creatures in the development of the relevant features and capacities. I would also claim that at most only a very few of our fellow creatures are capable of developing intellectual lives at all resembling ours. In sum, I think we have good empirical grounds for regarding ourselves as special.

I do not understand any of the features of consciousness I have pointed to well enough to know whether they should be regarded as occurring in all degrees of development or whether they are intrinsically "lumpy." That is, is it possible for a conscious creature to be just the slightest bit self-conscious, or is self-consciousness an all-or-nothing matter? It seems implausible to regard self-consciousness as strictly an all-or-nothing matter, but there may still be some initial threshold of development below which there can be no self-consciousness at all. In any event, to the extent that the features of consciousness I have pointed to do come in degrees, I would concede that different degrees of valuable consciousness may all produce valuable complexes with, of course, different degrees of value.

So there *may* be some intrinsic value in the intellectual life of a dog. It does not follow, however, that the dog in whose life there is some intrinsic value is as valuable as, or has the same "rights" as, a fully developed human being. Because there is more good to be achieved by "cultivating" the human, the human's needs and interests will tend to prevail.

So much for my argument about why, and to what extent, humans are special. Many environmentalists seem to think that claiming to be special can show only that we are arrogant or presumptuous. Paul Taylor has recently attempted to show that humans should not be regarded as special.[16] Taylor's argument is aimed specifically against the claim that we are special because we are "rational." That is not exactly my claim, but it is close enough to make what he says of interest here. His argument goes as follows: We cannot say humans are special just because they are more rational than other animals. After all, cheetahs are faster than other animals, and eagles see better. If humans are best at one thing, other species are best at other things. Nor can we say that being rational is better or more important than being fast or far-sighted. Being rational has been adaptive for humans and is valued by humans. But cheetahs and eagles have taken other routes and value other qualities.

This does not seem to be an argument at all against a position like mine. It may be a useful reminder that we cannot say simply, "Rationality is good because that's what we're good at," or "Rationality is valuable because that's what we value." But it says nothing against the possibility that there is some objective point of view from which rationality does actually have an importance which speed and acute vision do not.

Indeed, Taylor himself seems committed to the existence of such an objective point of view. He is one of those who believe that every creature has a good of its own. He instructs us that we should respect the goods of other creatures. But why? If the speed of the cheetah is good only from the cheetah's point of view, as Taylor suggests rationality is good only from our human-centered point of view, then why should we respect it at all? The cheetah's point of view is not ours. In telling us to respect the goods of other species, Taylor implicitly assumes both that there is a viewpoint "above" all the particular species-centered views and that we can adopt that viewpoint. If this synoptic viewpoint exists and is accessible to us, why should we not find when we adopt it that our capacities have just the special importance I claim?

My own view, of course, is that there is a synoptic viewpoint, and that it is accessible to us, and that when we adopt it we see not only that humans are specially important, but also why. What we see is that humans are necessary to the full realization of the "goods" of other species. The cheetah's speed is good, but it is not good in itself. It needs to be known by a subject who can know it and take pleasure in it in a sophisticated way. The cheetah does not value his speed in the required way. We can and should. This is the proper spelling out of the notion that every creature has "a good of its own."

One of the most striking articulations of the claim that humans are special comes from none other than Aldo Leopold, who is often thought of as concerned only to deflate human pretensions and put us in our place. In an essay inspired by a monument to the passenger pigeon, Leopold writes:

> For one species to mourn the death of another is a new thing under the sun. The Cro-Magnon who slew the last mammoth thought only of steaks. The sportsman who shot the last pigeon thought only of his prowess. The sailor who clubbed the last auk thought of nothing at all. But we, who have lost our pigeons, mourn the loss. Had the funeral been ours, the pigeons would hardly have mourned us. In this fact, rather than in Mr. DuPont's nylons or Mr. Vannevar Bush's bombs, lies objective evidence of our superiority over the beasts.[17]

X. On Not Avoiding the Problem, Again

I am now done with expounding my solution to the problem posed at the beginning of this essay. The solution, depending as it does on a Moorean view of the good, may seem implausible. Would it be better after all just to deny that we have noninstrumental duties of preservation?

Many philosophers who write about our relationship to nature observe that we have intuitions suggesting noninstrumental duties of preservation and conclude immediately that some theoretical underpinning for such duties must be found. That is not my position. One function of moral theory is to account for our intuitions; but another function is to test them. If we had a moral theory that was satisfactory, or even reasonably satisfactory, in other respects, but that failed to account for our supposed duties of preservation, then we would have strong reason for denying those duties.

In truth, however, we have no theory which is even remotely satisfactory in other respects. There was a time when philosophers treated as the central question of ethics the question, "What ought I to do with my life?" I think this is still the central question. But it is a question which neither the modern utilitarian nor the modern Kantian (and every popular theory is either a utilitarianism or a Kantianism) even attempts to answer.

The modern utilitarian takes as the good the satisfaction of people's preferences, whatever they may be. But this theory of the good does not specify adequately what we should aim at in life, for the simple reason that human preferences are not given exogenously. Any individual can to some extent control the content of her own preferences. She cannot revise them instantaneously by an act of will, but she can act in ways which, over time, will create or strengthen some preferences and extinguish or weaken others. Even more obviously, the general run of human preferences are in large part the result of prior human decisions, by families, societies, and polities; and present decisions by families, societies, and polities affect what people's preferences will be in the future. We are therefore inevitably confronted with the question of what our preferences ought to be. What ought we to try to bring about that we and others prefer? Aside from some observations about the desirability of preferences being, so far as possible, well informed and consistent, modern utilitarians have nothing to say about this. They cannot have anything to say about this without going beyond the claim that the good is the satisfaction of preferences. (Benthamite utilitarianism, with its claim that the good is a single, homogeneous stuff called "pleasure," is not subject to the same criticism. But there are other reasons why Benthamite utilitarianism is implausible. One cannot believe after reading the *Gorgias* and the *Philebus* that pleasure is the good; and one cannot believe after reading the *Nicomachean Ethics* that pleasure is homogeneous.)

The modern Kantian—and I use the term broadly, to include libertarians as well as liberals—is in no better state. His essential program is to carve out for each agent a sphere of autonomy. This may be done (as the libertarian does it) by purely negative constraints on how individuals may be treated; or it may involve (as it does for most liberals) positive prescriptions concerning such things as the distribution of resources. In either case, nothing is said about what anyone ought to do with the autonomy that has been secured for him.

If modern moral theory ignores the central question of morality—"What ought I to do with my life?"—then we need a new moral theory, or at least a different theory from the currently popular utilitarianisms and Kantianisms, for reasons which go far beyond the desire to justify preservationist intuitions on philosophical grounds. It seems to me that some theory of the good along the general lines of G. E. Moore's account will be necessary to answer the question of what one ought to do with one's life. It is a happy accident that such an account also allows us to explain duties of preservation which are otherwise inexplicable. (It isn't really an accident, of course. Any proper answer to the question of what one ought to do with one's life will have to address the general problem of man's place in the universe.)

217

XI. CONCLUDING REMARKS

This has been a long essay, a forest in which we have become preoccupied with a number of individual trees. I have two concluding remarks.

The first remark is by way of summary and emphasis. The main thing I set out to do was to construct a theory with two features. On the one hand, the theory should generate categorical duties of preservation—duties that do not depend on instrumental arguments, and, more particularly, duties that are independent of our changeable needs and preferences. On the other hand, the theory should be consistent with our belief that the existence of value depends on consciousness; the theory should avoid ascribing intrinsic value to natural objects indiscriminately. I have constructed such a theory. Whether I have made it plausible, or as plausible as the alternatives, the reader must judge.

The second remark is not central to the present essay, but it is important to a correct general understanding of our duties regarding nature. If my theory is right, human beings occupy a special role in the universe. We are essential to the realization of the most significant values, even those involving nonhuman nature. What that means is that all the standard instrumental arguments for preservation, if they are sufficiently general, retain their full force on my view. "Sufficiently general" instrumental arguments are those based on what is essential to human life and general well-being, as opposed to what is necessary to satisfy particular, possibly valueless, preferences. Thus, although I have been concerned to construct a noninstrumental argument for certain conclusions, I have definitely not attempted to displace instrumental arguments across the board. It may well be that the most important arguments in practice are still instrumental. That also accords with common intuitions.

NOTES

1. Cf. William Frankena, "Ethics and the Environment," and Peter Singer, "Not for Humans Only: The Place of Nonhumans in Environmental Issues," both in *Ethics and Problems of the 21st Century*, edited by K. E. Goodpaster and K. M. Sayre (Notre Dame: University of Notre Dame Press, 1979), pp. 3-20, 191-206; Henry Sidgwick, *Methods of Ethics*, bk. I, chap. 9, sec. 4 (New York: Dover Publications, 1966).
2. Before going on, let me put aside a possible distraction, the issue of animal pain. Almost everyone now agrees that we have at least one nature-regarding duty which is not to be explained in terms of human needs or preferences. I refer to the duty not to cause pain to other sentient creatures. The recognition of this duty does not solve, or even suggest a solution to, our larger problem. The duty not to cause pain does not begin to encompass all the

nature-regarding duties we seem to have. On the other hand, we can recognize a duty not to cause pain without seriously undermining the general notion that value depends on consciousness. In short, we can recognize a duty not to cause pain without thereby either accounting for our supposed duties of preservation or eliminating the apparent obstacle to accounting for them. A completely adequate account of why we should not cause pain would shed some light on the nature of the connection between value and consciousness; but we do not have a completely adequate account of why we should not cause pain. We do not have an adequate account of what pain is. In the present state of affairs, recognizing a duty not to cause pain leaves the problem of duties of preservation untouched. In what follows I shall ignore the duty not to cause pain, without meaning to deny it; and I shall not worry about how it fits in with the broader theory I shall describe. For a further discussion of these issues, see Elliott Sober, this volume.

3. Judith Jarvis Thomson, *Self-Defense and Rights* (Lindley Lecture) (Lawrence: University of Kansas, 1977).

4. For example, Paul Taylor, "The Ethics of Respect for Nature," *Environmental Ethics* 3 (1981): 197-218. Similar claims are that all living creatures have "interests," Kenneth Goodpaster, "On Moral Considerability," *Journal of Philosophy* 75 (1978): 308-25, or that every living creature possesses a "will," Albert Schweitzer, *Civilization and Ethics* (Part II of *The Philosophy of Civilization*), trans. C. T. Campion (London: A. & C. Black, 1929), pp. 246-47, quoted in Peter Singer, *Practical Ethics* (Cambridge: Cambridge University Press, 1979), p. 91.

5. Baird Callicott, this volume.

6. For Moore's treatment, see *Principia Ethica*, chap. VI (Cambridge: Cambridge University Press, 1903). The reader who knows his Moore may be thinking that Moore would admit some value in the Canyon all by itself. That is true, and it is a point I shall return to later on. Whatever Moore would have thought about the value of the Canyon by itself, it is certain that he regarded "organic unities" of the sort I have described as having a value which was not merely the sum of the values of their parts. It is equally certain that Moore thought these unities contributed much more to the overall value of the universe than did objects like the Canyon on its own (*Principia Ethica*, chap. III, sec. 50). In any event, I am not primarily interested in exegesis of Moore. Although I have taken a fundamental idea from him, and although I am indebted to him at many points, I shall develop his idea in my own way and for my own purposes, without attempting to specify where I am following him and where I am not.

7. Cf. Holmes Rolston III, "Values in Nature," *Environmental Ethics* 3 (1981): 113-28; Tom Regan, "The Nature and Possibility of an Environmental Ethic," *Environmental Ethics* 3 (1981): 19-34.

8. My suggestion concerning what it means to say the Canyon is worth knowing about parallels Moore's definition of the beautiful as "that of which the

admiring contemplation is good in itself." *Principia Ethica*, chap. VI, sec. 121.

9. Richard Routley, "Is There a Need for a New, an Environmental, Ethic?" *Proceedings of the XVth World Congress of Philosophy* 1 (1973): 203-209.

10. The points in the text about the last person argument can be reformulated in such a way as to undermine Moore's argument (against Sidgwick) for the proposition that inanimate objects may have intrinsic value, at least insofar as Moore suggests an argument and does not merely rely on his intuitions. See *Principia Ethica*, chap. III, sec. 50. Moore may have gone over to Sidgwick's side on this issue by the time he wrote *Ethics* (London: Oxford University Press, 1912), p. 153.

11. Cf. Elliott Sober, this volume.

12. Cf. Terry Leitzell, this volume.

13. Cf. Kenneth Simonsen, "The Value of Wildness," *Environmental Ethics* 3 (1981): 259-63; Aldo Leopold, "Goose Music," in *A Sand County Almanac: With Essays on Conservation from Round River* (New York: Ballantine Books, 1970), pp. 226-33.

14. Perhaps I should also point out that I am not backing off from the claim that species have intrinsic value only within certain complexes. Strictly speaking, the question I shall discuss concerns the comparative intrinsic value of complexes involving different species. It is much more convenient, however, to speak of the intrinsic value of species, and for the course of the discussion of priorities I shall do so.

15. I am also ignoring the sort of question that Slobodkin discusses in his very interesting essay. His concern, roughly, is how we should expend scarce resources to identify endangered species. Given our considerable ignorance about which species are in danger, this question may be much more important in practice than mine. See Slobodkin, this volume.

16. Paul Taylor, "The Ethics of Respect for Nature," *Environmental Ethics* 3 (1981): 197-218.

17. "On a Monument to the Pigeon," in *A Sand County Almanac*, p. 117.

Management
Considerations

Introduction to Part III

In dealing with the endangered species problem, it is tempting to assume a straightforward management model that has the following elements: (1) a scientific description of the current situation; (2) a statement of values and goals, ordered according to their priority; (3) a statement of policy objectives—a list compiled by comparing and noting disparities between (1) and (2); (4) a scientific research effort designed to obtain the information necessary to pursue the objectives efficiently; and (5) the development and dissemination of an instruction manual (a sort of recipe book) telling managers exactly how to proceed in whatever specific situations might arise.

But the chapters comprising this part of the book have little to do with this neat and tidy model, because the model assumes a level of knowledge about the present situation, a consensus on values and goals, and a research capability for gaining the necessary new information that far exceed current agreements and capabilities.

No reliable, let alone detailed, description of the current situation exists. Statements of current rates of extinction are unfortunately very speculative. Further, we have little knowledge (as opposed to speculation) as to where, geographically, the greatest effects on biological diversity are occurring. Nor is it known precisely which types of ecosystems are most threatened or what human activities entail the greatest threats.

Managers must face seemingly overwhelming problems with very little detailed data and even less generally accepted theory concerning the functioning of ecosystems. We do not completely understand species interactions or the levels of redundancy within ecosystems; consequently, we cannot be sure which management strategies will have the desired results.

Nor is there any generally agreed upon comprehensive analysis of the value of other species. For example, the rationale for the U.S. Endangered Species Act of 1973 states that "endangered" species of fish, wildlife, and plants are of esthetic, ecological, educational, historical, recreational, and scientific value to the nation and its people [Endangered Species Act of 1973, Sec. 2(a) (3)]. The largely alphabetical order indicates that the authors of the legislation despaired of finding some more principled way of ranking the various vague values that motivated the Act.

Even less agreement would exist if these general, abstract values were translated into specific proposals to set aside wetlands as opposed to forests, to protect individual species rather than ecosystems, or to concentrate on protecting some segment of the phylogenetic scale at the expense of others.

But the inapplicability of the ideal model need not be reason for despair. The absence of desirable information and broad agreement should not leave managers paralyzed. We do have at least enough information to rule out the assumption that every action is as good as the next. There are some middle-level scientific generalizations that can guide policy and there are some courses of action that serve more stated values than others. For example, we know in general which characteristics of species and their populations make them more or less susceptible to extinction. Thus, while we lack a general theory of speciation and extinction and detailed knowledge of most species' life histories, we do have a good idea of which species require special attention and why. The goal of this final section is to summarize what is known, to suggest preferred courses of action under less than ideal conditions, and to outline strategies for broadening the base of support for species preservation.

Lawrence Slobodkin illustrates this middle-level theorizing by outlining the generally accepted ecological insights likely to be most useful to environmental managers. Populations and individual organisms both have properties that make extinction more or less likely. An understanding of these properties and their dynamics should allow managers to predict which species are likely to be at risk in various situations and to concentrate their efforts where they are most needed.

He also points out that if human perturbations of ecosystems can be designed to mimic natural environmental alterations, organisms are more likely to be able to adapt behaviorally or physiologically; these are more promising adaptational strategies than genetic adaptation, which often proceeds too slowly to respond to anthropogenic alterations of the environment. Slobodkin suggests that with adequate care and forethought based upon what we do know about susceptibility to extinction, we could minimize human activities that clearly endanger the survival of other species.

Slobodkin's essay exploits available information to show how to avoid threats of extinction; Terry Leitzell's contribution addresses the practical question of how a manager can and should proceed, given that crucial pieces of information are often not available. He evaluates management strategies comparatively, describing which techniques are likely to work in different situations. His goal is twofold: first, to show how various strategies can serve a wide range of values even when no clear consensus guides the setting of priorities; and second, to argue that, when the scientific knowledge necessary to manipulate populations is unavailable, protection of large areas of habitat is the best strategy. If a species still exists in the wild, its essential needs are obviously being met by the natural environment. If humans are ignorant of the species' needs and life cycles, the best course of action is to reduce anthropogenic pressures on the natural ecosystems that form its habitat.

The lesson of Slobodkin's and Leitzell's essays, taken together, is that good and sensitive environmental management will make use of available scientific knowledge while recognizing the extent of our ignorance. Human intervention in ecosystems is, after all, the root of the current wave of extinctions. In the absence of knowledge adequate to intervene for the protection of species, the best course is often to set aside as large a preserve as possible, allowing natural processes not properly understood by humans to serve the needs of threatened species.

Robert Carlton looks at strategies to protect species in a larger, societal perspective. He argues that current U.S. legislation is skewed toward restrictions and prohibitions, and that this approach is both inequitable and ineffective. Legal procedures that penalize individuals when endangered species appear on their property or in their workplace will not encourage wide participation in measures for species preservation. He recommends a more positive approach; developing incentives for cooperation is more likely to achieve success than enforcing restrictions.

This last part of the book, then, tackles the question of what to do when one does not know what to do. The proper response, it suggests, is to make use of the knowledge that is available, to seek aggressively that which is not, and to design current actions to do as little harm as possible. Protecting large areas of habitat from human encroachment and allowing nature to continue its processes without human interference is often the best strategy, given current levels of ignorance and perennial disagreements about goals and objectives. The conclusion is a humble one, but then again, humility may be, above all, the virtue we as a species are most in need of acquiring.

9

On the Susceptibility of
Different Species to Extinction:
Elementary Instructions for
Owners of a World

LAWRENCE B. SLOBODKIN

I. INTRODUCTION

Extinctions have occurred before and will again—even in a world free of anthropogenic perturbations. In a similar sense, human deaths have occurred before and will again, even under the best of social systems. To act in such a way as to cause a death or an extinction, however, is not excused by these historical precedents.

Other chapters in this volume focus on the list of species that are known to be endangered and the managerial options available to circumvent the danger. They also provide examples of past extinctions. I will not duplicate these efforts, nor will I consider the problem of the consequences of extinction, except to note that the potential consequences of a species extinction may extend well beyond the species itself. At the most concrete level this statement refers to ecological changes that may ramify through a system following an extinction. At another level arise deep philosophical and ethical questions, which lie beyond the normal boundaries of scientific discourse.

I will focus on a slightly different problem. What can we say about the susceptibility of species to extinction? This is divisible into questions such as: what are the properties that make certain species relatively immune to the threat of extinction? When the environment is changed, what kinds of species are most liable to become extinct?

The chapter is written in a didactic tone, inspired by the set of introductory operating instructions that might come with a new refrigerator or lawn mower. Such devices will work reasonably well, so long as one is reasonably clear on what functions they are to serve, how to tell when they are functioning properly, and when to call a repairman. One need not understand metallurgy, thermodynamics, electronics, or cybernetics to be able to run a refrigerator, but it is important to know when it sounds wrong.

The most basic kind of instruction booklet tells us what we need to know to avoid immediate disaster. It also tells us what danger signs to look for. If these signs appear, call in someone who understands the system. In that spirit, I provide a brief guide to what might go wrong with ecological systems, preceded by a brief rationale for my conclusions.

My assertions are often condensations of my past thirty years of reading in ecology and evolution. If they appear to you to be outlandish, incorrect, or dubious, please refer to recently published summaries, which taken together provide an introduction to the literature.[1] These works should give enough of a general sense of ecology to permit evaluation of my summary remarks. In particular, they will lead to empirical examples.

In the spring of 1950, I crashed my bicycle in the driveway of the Osborn Memorial Laboratory at Yale. Three passing faculty members hurried to my aid. A philosopher and physicist, Hans Margenau, checked my bicycle, and the developmental biologists, Edgar Boell and John S. Nicholas, checked me. I then realized that under my particular circumstances an advanced medical student and a garage mechanic would have been more comforting. In this spirit I have not recommended texts at the forefront of controversy in the field. In fact, my intention in this article is to avoid any novelty or speculation. It is sometimes difficult for nonexperts in a field to tell the difference between long-accepted notions, on which complete reliance can be placed, and the current research concern of a particular author. New theoretical speculation and even new data are of necessity subject to revision and testing. There is a danger that untested material may be too readily believed by readers who are not familiar with the tentative and qualified nature of many ecological assertions. Just as a first-aid manual should avoid explanations of frontline experimental medicine, in the interest of everyone's well-being, so I have avoided controversial assertions, even when I believe in them rather passionately.

I hope that this chapter will be read by nonbiologists. It is therefore my intention to indicate the kinds of practical ecological problems that are likely to require decisions by nonecologists. These occur on a scale ranging from those that require the attention of politicians and planning supervisors to those faced by voters and backyard gardeners. Even for these readers I will only provide general insights into when problems are serious enough to require professional advice.

Given the complex needs of human societies, circumstances may arise in which it is necessary to extinguish populations or even species. In many cases, however, extinction occurs because of ignorance or carelessness. Are there biological warnings that may aid us in avoiding thoughtless extinctions? Is there any general way of assessing the likelihood that a particular species might become extinct? How are the dangerous situations to be recognized?

We cannot answer these questions by cataloguing the separate properties of the 10^6 species of organisms that now exist or the 10^9 species that may have existed. We may, however, be able to specify the biological and environmental properties that either provide a certain amount of protection against extinction or, conversely, make certain populations, species, or ecosystems susceptible to the danger of extinction.

Serious ecological problems may literally be infinite in number.[2] To pretend to provide a solution for some particular ecological problem by use of a general formula would be a serious disservice. The harm done by amateur psychoanalysts and political philosophers is mild compared to what might be done if the ecological world were to be managed by someone who had read one pamphlet and thereby felt sufficiently confident to take action. However, a guide to recognizing problems that require expert opinion may be of value in preventing accidental extinction.

II. The Death of Individuals and the Death of Populations

Extinction is one kind of death. It is the death of a population of organisms, which may coincide with the death of a subspecies or species. Death can be formally defined as the cessation of existence of some definable entity. Less formally, the meaning of death varies with the entity considered and the reference point of the observer. "The Death of the Wonderful One Horse Shay" is comic, while the death of one's children is the epitome of tragedy. Some kinds of death are inevitable, given our knowledge of natural law. Among these are the death of a fire on limited fuel and the death of most individual organisms. Death is not inevitable on all ecological levels, however. There are clear reasons why I expect all individual organisms eventually to die, but these same reasons cannot simply be extrapolated to the population level. No natural law requires that populations or species should become extinct from anything other than catastrophic environmental changes.

228

I will first consider the inevitability of individual death, then the lack of inevitability of extinction, and lastly the meaning of "catastrophe" in this context. The managerial procedures for avoiding extinction of valued species involve the capacity to avoid catastrophe as it will be defined here.

In one sense there is obviously a genetic program for death, but its mechanism varies among organisms.[3] The death of individual organisms occurs either because vital cells die or vital organs are irreparably damaged. When recurrent and predictable environmental events make living impossible, death may be programmed into the developmental pattern of the organisms. Annual plants are programmed to die after setting seeds. Most anadromous fishes die after breeding, as do many molluscs, annelids, and insects. In mayflies, in some male salmon, and in some of the squids, the gut simply atrophies prior to the onset of breeding as if to underline the inevitability of individual death.

The inevitability of a salmon's death is not predictable on the grounds of general biochemistry or cellular biology, except in an ecological and evolutionary context. Most organisms contain organs and structures that are not naturally replaced, such as the teeth of mammals and wings of insects. There is, however, a real possibility that those organisms that consist only of parts that can be replaced or repaired may be immortal in principle, if not in fact.

In essentially all organisms, regardless of age or of the inevitability of death in the immediate future, there are some cells actively replacing or maintaining their subcellular parts. In our central nervous system there exist cells that cease dividing prior to the fifth year of life but that may then persist and function perfectly for another century. Other cells continue to undergo cell division, even immediately prior to death. Typically these are the cells that, by being abraded away, form protective coverings for the others. The cells on root tips of 4000-year-old bristle cone pines ease the passage of the roots through the soil, in just the same way as the root tip cells of suburban tomatoes function. As salmon swim upstream, epithelial cells continue to divide on their skin, as if they were not aware of the approaching end.

If the biochemical potential for life is so persistent, why is death inevitable? The answer relates to evolutionary mechanisms, rather than simply to biochemistry.

Some organisms that live in unchanging worlds or in worlds in which the pattern of change is reasonably predictable, even though the changes may be very drastic indeed, may live a very long time as individuals. These kinds of organisms also evolve very slowly. The blue-green algae that dry to a tar-colored stain on boulders between wettings by the surf, the sponges and coral heads that are known to have persisted for centuries, and the gravestone lichens that have not much changed their size in centuries are of this type.

For most organisms, however, the world offers accidents that matter, so that to delay too long before being ready to reproduce may result in no reproduction at all. To advance the onset of reproduction, however, may generate weaknesses and susceptibility to other sources of mortality. From the standpoint of natural selection, organisms that have a genetic propensity to transmit genes most effectively into the long-term future of their population will have, in the long run, an evolutionary advantage over other members of their population. They are the "fittest" that nature "selects." This is true even if the process of gene transmission is detrimental to individual longevity.

Why not do both? Why not be strong enough to withstand the environment and, simultaneously, fecund enough to meet the criteria of evolution? The answer is not perfectly clear. Resources are apparently limited, and within the individual an assignment of resources to any particular function limits the resources for other assignments. This results in "evolutionary constraints." For example, a protective shell against predators sets up a series of costs in making the armor and carrying it about. These costs probably interfere with speed, but also with reproduction and perhaps other things. In short, no single organism can optimize its capacity to respond to all possible perturbations.[4]

Notice that my arguments for the inevitability of the death of organisms are based on the evolutionary need for reproduction. This process replaces a senescent individual with one or more new ones. In a metaphorical sense it is for the sake of these replacement individuals that older individuals must die, and because of these replacements the average age of the individuals in a population need never grow older. While populations of a particular type may have very long histories, they do not age, i.e., undergo senescence, in the way that individual organisms do. In principle, populations of organisms need never die.

Even accepting the above argument, I have not yet eliminated the possibility that extinction of populations is inevitable for other reasons. The two possible mechanisms for the death of populations are random fluctuations and major environmental perturbation.

It is easily demonstrated that any sequence of numbers, such that the next number in the sequence is a random function of the previous number without regard for their magnitude, will wander away from any constant value. This is termed a "random walk" or "drunkard's walk." Imagine a "drunkard's walk" in terms of a person actually walking down a rectangular platform, always moving down the platform, but lurching at random to the right or left, without any regard to proximity to the platform edge. If the platform is long enough, the walker is essentially certain to fall off, even if it is a very wide platform. Therefore, extinction of populations would be inevitable if the changes in abundances of individuals in populations could be rigorously modeled as a "random walk." This type of random walk would occur if the physiological state of the individuals in the population, and of all of its predators and prey, was unchanging and independent of their abundance. I could then define particular years or other time intervals as being better or worse for the population. At better times the environment would suit the organisms' physiological state in such a way as to permit the population to increase. At worse times the population would decrease. If the distribution of these times was itself random, the population would then wander in size from year to year. Since there is an absorbing barrier at zero for a self-reproducing collection of mortal organisms (i.e., extinction is irreversible), we could expect that extinctions of populations would be inevitable under this model.

In general, when local populations are considerably more abundant than their average value, it can be demonstrated that sooner or later their food supply, or nesting places, or some other environmental requirement will become lower than average in quality or quantity for the average individual in the population as a consequence of among-organism competition. Conversely, when populations are reduced, competition for these resources among members of the population is reduced. As a matter of fact, for all populations that have been examined with sufficient care, it has been possible to show that, for most populations studied, reproductive rates vary inversely with local population density and that mortality rates vary directly with population density. These rate changes may be mediated through any of a broad variety of mechanisms, but the important conclusion for our purpose is that populations more closely approximate feedback systems than random walk systems, so that drunkard's walk extinction cannot be considered as inevitable.[5]

If the death of populations does not come about from senescence, nor from random walk theory, how do we account for the fact that the paleontological record indicates that approximately one hundred times more species are extinct than have survived?[6] If extinction is not inevitable in principle, why have most species become extinct? We are left with catastrophic environmental perturbation as the main cause of species extinction.

III. How Organisms Respond to Perturbations

Environmental circumstances change for all populations, even those in the abyssal depths and other situations that are usually called "constant." There is normally a spectrum of kinds, rates, and magnitudes of changes in the environment of any organism to which it has a spectrum of suitable response mechanisms, such that these events are not disturbances in any serious sense. Any anthropogenic event that mimics a component of this spectrum will be responded to in a normal way. Such an event is not "seen" as a change by the organism. It follows that a particular event may not be of any significance to one kind of organism but may be of great importance to another.

Within a single natural population, individual organisms differ in age, nutritional state, and in their capacity to either resist deleterious environmental circumstances, or to take advantage of favorable ones. The differences among organisms can be attributed to their individual past histories and to their genetic endowments, their genotype. Differences among organisms in their response to mild environmental changes are usually attributable to obvious physiological differences, rather than to genotypic differences. Genotypic differences in the capacity to survive and reproduce are generally only demonstrable in response to severe environmental changes. These differences may result in increasing or decreasing of particular genes in the population, from one generation to the next. When such changes of the relative abundance of genes, the gene's frequency, have occurred, the population is said to have evolved. These gene frequency changes may be expected to have altered the population in such a way as to reduce the deleterious effect of whatever environmental stress caused the gene frequency changes in the first place. The greater a population's capacity to minimize numerical changes and genetic changes as a consequence of a particular environmental event, the less the danger that such an event will cause extinction. Any disturbance to the organisms in a population is responded to by a series of simultaneously activated mechanisms, ranging from immediate behavior to genotype-specific alterations in survivorship and fecundity patterns. Each of these responses has its own temporal properties. Gene frequency change in a population takes longer, under most circumstances, than individual physiological changes. Therefore, if a perturbation is responded to appropriately on a behavioral or physiological level, it may cease to matter long before any major change in gene frequency could occur. It is in this sense that appropriate, rapid responses protect organisms against the necessity of making slower, less reversible responses.[7] If the environmental change is within the normal range of events to which a particular kind of organism can respond, I expect the behavioral responses to be such as to minimize the need for physiological response, and the physiological responses to minimize the changes in mortality and fecundity, and so on.

Even if the behavioral and physiological responses have been inadequate, so that mortality rates have increased, the possibility remains that reducing population size, and so reducing intrapopulation competition, will enhance the surviving organisms' resistance to the environmental change that caused the damage and to other environmental problems as well. This idea, described by the term "compensation," has figured prominently in the legal disputes over the effect of power plants on fisheries. It is also used in fisheries management situations. While "compensation" certainly exists for some populations and some environmental events, it must be considered with great care. In particular, it cannot be used as a shibboleth to justify assertions about the innocuous character of particular events.

The greater the sensitivity of organisms to the density of their own species, the greater their capacity to recover from population depletion. Members of greatly depleted populations may have increased survival and fecundity as a consequence of low intra-specific competition. While the individual organisms of these small populations are at their physiological best, further depletion of the population may result in extinction.[8] Forms that can become abiotic spores and wait for the world to become reasonable again cannot be extinguished—nor need they evolve much. However, if the rate and magnitude of detrimental environmental change are great enough to exceed the response capacities of individuals, and ultimately of populations, species may become extinct.

Speciation and extinction are both accelerated in groups with restricted physiological and behavioral buffers between their germplasm and environmental disturbance. Evolutionary changes involve changes in gene frequency. But gene frequency changes occur only if the behavioral and physiological mechanisms that normally buffer the population have failed. Therefore the same circumstances that enhance the probability of extinction also accelerate the rate of evolution. I believe that it will be shown that those groups, such as the Hawaiian Drosophila, that have speciated very rapidly also have a very high probability that any one species will become extinct in the future.[9]

Notice that some insect populations exhibit seasonal changes in gene frequency that seem to be adaptive in themselves. These gene frequency changes tend to be seasonally cyclic and are not the kind of genetic changes being discussed.[10]

233

IV. LOCAL AND GLOBAL EXTINCTIONS

The term "extinction" has two meanings: "local" extinction refers to a particular population and "global" extinction refers to the entire species. Local populations may disappear, either temporarily or permanently, without necessarily implying extinction or even a dangerous situation that might lead to extinction. On the other hand, a species is composed of the sum of its populations, so that elimination of populations may imply danger to the species as a whole. The global extinction of a species occurs at the moment of extinction of the last reproducing unit of the last population.

Susceptibility to extinction is dependent in part on how the species is subdivided into populations. The populations of a species can be more or less spatially separated from one another, in the sense that the region between the habitats of the particular populations may contain few or no specimens of the species in question.

Some species normally occur in a small number of populations, each of which constitutes a more or less discrete entity. The large migratory game animal herds may exemplify this arrangement as do the populations of highly social and relatively rare primates. There may have been only one or two gigantic populations of passenger pigeons. Other species occur as one or more networks of populations, such that within each network local subgroups may fluctuate in abundance, send propagules to other subgroups, and receive new members. This is the pattern for many insect populations.

The significance of local population reduction or elimination is very different in these two extreme cases. A species subdivided into a large number of semiautonomous populations over a wide geographic range is less susceptible to extinction than a species consisting of only one population.

Anadromous fishes are a convenient intermediate example: such fish mingle freely in the marine phase of their existence, but then separate into separate runs, located in very specific streams, when breeding is to occur. From the standpoint of breeding populations, each stream's population must be considered individually, while from the standpoint of managing a marine fishery the subunits may be pooled into a kind of superpopulation. The extinction of a stream's population and that of an entire marine superpopulation obviously differ in significance.

It might seem that if there are many populations the loss of one does not matter. This view does not take into consideration the possible ramifications of that population's ecological role. This role may include being central to the local ecology (rabbits as a controlling herbivore in England or wolves as a controlling predator in North America), or having important amenity value (deer at the shrine at Nara or ravens at the Tower of London).

V. The Causes of Rarity

Broad geographic distribution over a range of habitat types provides some insurance against extinction, but absolute abundance does not. Just before becoming extinct a species is very rare. Rarity, however, does not always imply incipient extinction, nor is abundance a universal measure of safety from extinction. Very rare organisms may be rare for a very long time and in many places or they may be the last vestige of what was once common. Passenger pigeons may have been the most common of pigeons at the time of Audubon, but they are now gone. Bison herds contained so many individuals that they took days to pass a single point, but the extinction of bison was just narrowly avoided. The true Mormon cricket is believed to be extinct although they were enormously common in the nineteenth century.[11]

Rarity can be explained either by ecological peculiarities or by the diminishing of a population. It is important to be able to tell the difference. In the first case, the species will probably persist if the environment does not change. In the second case it will be eliminated unless the environment does change.

Persistent rarity can arise as a consequence of the fact that all animals either live directly on the energy-rich products of plant photosynthesis or live on other animals. The direct feeders on plants, the herbivores, use up most of their energy in metabolic processes. If we assume one thousand calories of photosynthetically fixed energy on one square meter of ground in one year, a herbivore might take approximately 200 calories from this area during the year, without destroying the plants entirely. A carnivore feeding on this herbivore can get around 20 calories per year from the same area without destroying its prey, and an animal feeding on the carnivore might get one or two calories from the same area. In this sense, the available energy per unit area of earth's surface becomes more "dilute" as we pass from plants to herbivores to carnivores. We never expect wolves to be as common as deer.

High-level predators also tend to be quite large, permitting unequivocal outcomes of encounters between them and their prey. Large size is associated with rarity, other things being equal. For both reasons, high-level predators tend to be rare relative to animals at lower trophic levels. Animal parasites are apparent exceptions. In one sense they are carnivores, but they are usually very small. Also, predators which hunt in packs (e.g., wild dogs) may be smaller than their prey.

If organisms have anabiotic stages, perhaps combined with a high capacity to disperse, their local rarity may not be a problem—unless the areas they occupy are eliminated or severely reduced. Organisms of this kind—for example, some blue-green algae, rotifers, and organisms living in hot springs and brine pools—may be rare but persistent for millions of years. Organisms such as mushrooms or plants with persistent seeds may show an illusory appearance of rarity, based on our inability to census the tiny spores or dormant seeds. Most of their populations are invisible to us.

If organisms that do not meet any of these conditions are rare, or if one knows that population sizes are declining, extinction is probably an immediate danger. Larger organisms tend to be relatively rare and also relatively more liable to extinction.

VI. Is Extinction Due to Evolutionary Momentum?

Some kinds of organisms are occasionally referred to in the press as "living fossils." This may mean that they are among a small number of surviving species in a taxonomic group that once was considerably richer in species, for example, the chambered nautilus, cephalocarids, and lampshells. Sometimes the term refers to organisms that have not changed over a long period, such as horseshoe crabs, sequoia trees, and opossums.

The term has an air of impending doom, as if the organisms were living on time borrowed from some pantheistic natural bank or as if they were decadent inbred natural aristocrats. As a matter of fact, brachiopods and opossums are extremely unlikely to become extinct. They meet most of the criteria for persistence listed above. In addition, they live in environments that are so variable from year to year and day to day that it is very unlikely that tomorrow will confront them with anything that they have not survived in the past.

Another category of living fossils is in much more severe danger. These are organisms that are adapted only to habitats that are themselves threatened and are in danger of disappearing completely or of being drastically modified.

In general, evolution operates by maximizing fitness of organisms in their specific environmental context. An organism in the center of a forest evolves for the circumstances of the forest. Organisms do not generally have the intellect or physical vantage point to view the world from a perspective other than their immediate surroundings. If drastically different adaptations are required for living in the forest compared with

236

living in the surrounding and encroaching farmlands, there is no way for a mid-forest organism to take this into account until the forest has become so small that there is no mid-forest.[12]

Green forests may not be in danger of vanishing, but natural mixed hardwood forests in which logs are removed by fire and rot rather than by trucks may certainly be made to vanish. Humans can take a broader view and maintain entire environments, within which other organisms may be able to use their repertoire of response mechanisms.

VII. Reintroduction—The Repairing of Local Extinction

So far I have indicated some of the properties that make particular species more or less susceptible to extinction. Among the signs indicating that a species is currently endangered is the local extinction of one or more of its component populations. The loss of a single population of a species that has several populations may not be permanent under all circumstances—sometimes, if it is done soon enough and with sufficient care, the population can be reestablished. Attempts at reintroduction are a standard managerial procedure. Unfortunately, more is usually involved here than simply transferring organisms from one region to another.

Three phenomena make reintroduction difficult. The first is that many organisms, particularly mammals and birds, are developmentally integrated into their local environments and cannot be expected to survive as well after being displaced. The reintroduction of raptors has required careful supervision and training of individual birds.[13]

Second, populations of particular species that have been subjected to different selective regimes in different regions may differ in gene frequency.[14] A reintroduced organism or group of organisms is likely to be genetically inappropriate to the region of reintroduction. In such a case the reintroduced animals are likely to be subjected to a severe selective regime, and this in turn may result in extinction of the reintroduced population. Notice that this type of extinction might in one sense be considered "natural," since the natural pattern of extinction can often be thought of as a result of inability to meet environmental demands—that is, inability to evolve sufficiently rapidly or effectively. Obviously, in the case of a reintroduction, this appearance of naturalness is specious.

The third barrier to reintroduction of a population is related to the fact that the initial elimination is likely to have provided an opportunity for competing species to occupy the ecological space originally occupied by the extinguished population. There is strong reason to believe that this may have happened in the Peruvian current fish population, in which the gross reduction in the anchoveta population has resulted in a restructuring of the local food web. The anchoveta's role in the food web has now been taken by herbivorous crustacean plankton,[15] and no attempt at anchoveta reintroduction is likely to be successful. This case also demonstrates that competitors are not necessarily as valuable as the species they replace. There is no prospect of a practical zooplankton fishery that could occupy the enormous empty anchoveta processing plants.

VIII. Buffers Against Extinction— Our Instruction Manual

These considerations suggest that it is possible to examine the potential danger of extinction of particular species as a function of their life history, distribution pattern, abundance, and competitive relationships. We have derived a sense of how some species are better adapted to avoid extinction than others. Declining abundance is a signal for immediate concern, but it should be possible to be aware of earlier danger signals. Knowing the danger signals would essentially constitute our set of elementary directions for determining which species we do not need to worry about and which species must be watched more carefully.

Susceptibility to extinction differs among species as a consequence of specific features of their biology. The likelihood of extinction of a particular species is minimal:

— If each of its populations is numerically large, so that deaths due to accidents will still leave some survivors.

— If it has highly resistant life stages (e.g., spores, dormant seeds, etc.), so that even the destruction of all the organisms of one life stage will leave a reservoir of survivors from another. (I am not aware of any endangered species of mushroom.)

— If it has a continual or at least a very long breeding season, so that short periods of abnormal environmental conditions do not eliminate an entire year's breeding.

— If adults survive through many breeding cycles, so that failure to reproduce in one season or year does not mean that an organism will never breed.

— If migratory rates between populations are relatively high, so that even if a local population is eliminated, it has a reasonable likelihood of being reestablished. (Consider garden weeds in this context.)

— If it is not subject to excess interspecific competition. By this is meant interspecific competition so severe that a drastic reduction in abundance or range would leave no "ecological space" for reexpansion, or reintroduction. Under severe competition, available space and resources are immediately encroached upon by other species.

Probably no single species possesses all these properties, although some organisms come close. Against a sufficiently severe perturbation, even all together will not avail.

IX. Conclusions

I hope that I have demonstrated that we have at least the beginnings of an understanding of ecological processes—an understanding that allows us to discuss the problem of which kinds of species are likely to become extinct and which kinds of anthropogenic changes tend to endanger species survival. This discussion was based on a set of physiological, ecological, and evolutionary generalities. Certain managerial recommendations are implicit in the above discussion:

— It is possible to assess the likelihood of extinction on the basis of biological properties, prior to a species actually facing immediate danger.
— Natural habitat preservation is preferable to preservation of species in zoos or botanical gardens.

Clearly, zoos and gardens are of enormous value as experimental resources, pedagogic devices, and aesthetic objects in their own right. They also have demonstrated their invaluable role as the last resort in combating extinction. Nevertheless, they are less certain as devices for species protection than is preservation of natural habitat.

Organisms can be more readily preserved in natural, or nearly natural, habitats than in artificial ones, because the latter require a much higher level of managerial effort and expertise. A natural preserve area may often be successful despite our ignorance and ineptitude, which is not true of zoos and gardens. A zoo or botanical garden is, by definition, an area in which organisms are cared for and protected by humans. This assumes that the caretakers have a reasonably complete understanding of the organisms' needs. In a natural habitat, organisms that have successfully met their own needs are permitted to continue to do so. It is more likely that a zookeeper or gardener will fail to notice or take account of an organism's needs than that the organism itself will fail.

Even in the best of practical circumstances, zoos and gardens are uncertain guardians of endangered species. We expect organisms to evolve in response to local circumstances. Long sojourn in a zoo or a garden may so domesticate any organism as to make its survival in nature impossible. Zoos and gardens are also subject to political and administrative pressures through which they may be reduced or even eliminated. There is clearly a spectrum of possible arrangements between preserves and zoos, in which the zoo or garden requires maximum human interference while a sufficiently large nature preserve may require almost none. A large number of plant and animal species cannot be maintained in captivity, and most of the ecologically important species are not the kind of organisms that appeal to zookeepers and gardeners. Soil bacteria, blue-green algae, and insects have a vital role in the maintenance of a natural habitat and its inhabitants, but are curiously uninteresting in zoos and gardens.

A sufficiently large natural population usually contains at least a few individuals carrying rare genes that may prove of value if drastic environmental changes occur. One of the problems in conservation of natural endangered populations is that the reduction in population size that is necessary in relatively small reserves or in zoos may lead to a loss of genetic variation in the population. This problem is also of importance for domesticated plants and animals, in which market requirements and considerations of productivity maximization often require genetically homogeneous populations of organisms. One procedure for alleviating this problem consists of maintaining gene banks, populations of domesticated species kept explicitly for their genetic variability. Deliberate attempts to maintain genetic variability share some of the problems of the maintenance of zoos and gardens, and, in addition, it is not quite clear how to utilize the resources in a gene bank. Most genetic variation stored in a gene bank is of no immediate utility, even for domestic species, and it is not quite obvious how one would organize and manage a gene bank for wild species.

Management aimed at only certain aspects of an ecosystem is generally more difficult and dangerous than management aimed at preserving the entire habitat. The more intimately manipulative the management process, the greater the approximation to the dangers of a zoo.

— Multiple preserves are preferable to only one. If natural or anthropogenic catastrophe destroys the sole refuge of an endangered species it will become extinct, but if there are several refuges it will only become more severely endangered.

— Large populations are not necessarily preferable to small ones, since I have demonstrated that abundance does not guarantee survival. This may be of help in planning ecological programs in a more deliberate fashion than the current mode of late response.

— Notice that the evolutionary process teaches organisms while it tests them, in the sense that the best guarantee of a population's surviving some future event is its having survived a similar event in the past. In fact, the meaning of the word "similar" in the previous sentence is the core of most conservationist lawsuits. Typically, the defendants in such suits try to demonstrate that their alterations of the environment do not differ in kind, or in magnitude, from natural changes that have occurred in the relatively recent past. The plaintiffs attempt to prove the novelty of the disturbance. The study of normal patterns of environmental disturbances and fluctuations is, therefore, almost prerequisite to setting and implementing environmental policy.

The overall brief conclusion is that an ecological approach to environmental management, drawing on what we know of evolution, species interactions, and life history strategies, will in general lead to more certain and useful procedures than thinking of particular species as if they were isolated specimens in collections. We need not wait until a species is "endangered" before considering its susceptibility to extinction.[16]

NOTES

1. M. E. Soule and B. A. Wilcox, *Conservation Biology: An Evolutionary-Ecological Perspective* (Sunderland, Mass.: Sinauer Associates, 1978); M. E. Soule and G. H. Herzog, *Conservation and Evolution* (New York: Cambridge University Press, 1981); R. E. Ricklefs, *Ecology* (Newton, Mass.: Chiron Press, 1973); G. E. Hutchinson, *An Introduction to Population Ecology* (New Haven, Conn.: Yale University Press, 1978); John Harper, *Population Biology of Plants* (New York: Academic Press, 1977); and L. B. Slobodkin, *Growth and Regulation of Animal Populations*, second enlarged edition (New York: Dover, 1980).
2. L. B. Slobodkin, D. B. Botkin, B. Maguire, Jr., B. Moore III, and H. Morowitz, "On the Epistemology of Ecosystem Analysis," in *Estuarine Perspectives*, edited by V. S. Kennedy (New York: Academic Press, 1980), pp. 497-508.
3. G. C. Williams, "Pleiotropy, Natural Selection, and the Evolution of Senescence," *Evolution* 11 (1957): 398-411; and G. C. Williams, *Adaptation and Natural Selection* (Princeton, N.J.: Princeton University Press, 1966).
4. T. H. Clutton-Brock and P. H. Harvey, "Comparison and Adaptation," *Proceedings of the Royal Society London, Series B* 205 (1979): 547-651; G. Bateson, *Mind and Nature: A Necessary Unity* (New York: Dutton, 1979); L. B. Slobodkin and A. Rapoport, "An Optimal Strategy of Evolution," *Quarterly Review of Biology* 49 (1974): 181-200; C. R. Townsend and P. Calow, *Physiological Ecology: An Evolutionary Approach to Resource Use* (Sunderland, Mass.: Sinauer Associates, 1981); and H. Dingle and J. P. Hegmann, *Evolution and Genetics of Life Histories* (New York: Springer-Verlag, 1982). Each of these works discusses, from different stand-

points, the problems associated with simultaneous optimizations of the properties of organisms.

5. N. G. Hairston, F. Smith, and L. B. Slobodkin, "Community Structure, Population Control and Competition," *American Naturalist* 94 (1960): 421-25; and L. B. Slobodkin, F. Smith, and N. G. Hairston, "Regulation in Terrestrial Ecosystems, and the Implied Balance of Nature," *American Naturalist* 101 (1967): 109-24.

6. F. S. Raup and S. M. Stanley, *Principles of Paleontology* (San Francisco: W. H. Freeman, 1971).

7. L. B. Slobodkin, "Toward a Predictive Theory of Evolution," in *Population Biology and Evolution*, edited by R. Lewontin (Syracuse, N.Y.: Syracuse University Press, 1968), pp. 187-205; Slobodkin and Rapoport, "Optimal Strategy," and H. C. Plotkin and F. J. Odling-Snee, "A Multiple-Level Model of Evolution and Its Implications for Sociobiology," *Behavioral and Brain Sciences* 4 (1981): 225-35.

8. L. B. Slobodkin, "Summary and Discussion of the Symposium," in *Fish Stocks and Recruitment*, B. B. Parrish, ed., Rapports et Procès-Verbaux des Réunions Conseil International pour l'Exploration de la Mer, 164 (1970): 7-14. Also J. A. Gulland, "Can a Study of Stock and Recruitment Aid Management Decisions?" ibid.: 368-72.

9. H. L. Carson, D. E. Hardy, H. T. Spieth, and W. S. Stone, "The Evolutionary Biology of the Hawaiian Drosophilidae," in *Essays in Evolution and Genetics in Honor of Theodosius Dobzhansky*, edited by M. Hecht and W. C. Steere (New York: Appleton-Century-Crofts, 1970), pp. 437-543.

10. Th. Dobzhansky, "Genetics of Natural Populations, XVI. Altitudinal and Seasonal Changes Produced by Natural Selection in Certain Populations of *Drosophila pseudoobscura* and *Drosophila persimilis*," *Genetics* 33 (1948): 158-76; and N. W. Timofeeff-Ressovsky, "Zur Analyse des Polymorphismus bei *Adalia bipunctata* L.," *Biologische Zentralblatt* 60 (1940): 130-37.

11. B. P. Uvarov, "Problems of Insect Physiology in Developing Countries," *Journal of Applied Ecology* (1964): 159-68.

12. Thomas E. Lovejoy, this volume.

13. T. J. Cade, "Plans for Managing the Survival of the Peregrine Falcon," Raptor Research Foundation Raptor Research Reports 3 (1974): 89-104; and T. J. Cade, "The Husbandry of Falcons for Return to the Wild," *International Zoo Yearbook* 20 (1980): 23-35. Also see various issues of *The Peregrine Fund Newsletter* of the Cornell Laboratory of Ornithology.

14. D. Futuyma, *Evolutionary Biology* (Sunderland, Mass.: Sinauer Associates, 1979), pp. 189-261.

15. J. Walsh, "A Carbon Budget for Overfishing Off Peru," *Nature* 290 (1981): 300-304.

16. See Terry L. Leitzell, this volume.

10

Species Protection and
Management Decisions in
an Uncertain World

TERRY L. LEITZELL

The last fifteen years have been characterized by rising concern in the developed countries for the preservation of species, the enactment of far-reaching legislation in the United States, and the entry into force of an international treaty designed to prevent, or at least control, trade in endangered species.[1] However, development interests, perhaps most strongly in the Third World, charge forward to improve the economy and to raise standards of living. Environmental groups continue efforts to strengthen current laws and to prevent extinction of certain species. The cacaphony of controversy can be deafening.

Government managers struggle constantly to find solutions that are acceptable to the hotly competing interest groups. They search for ways to allow rational development, while protecting and preserving the environment. They are often criticized for weak compromises and unproductive solutions. Yet, in many cases, it is impossible for the manager to construct a decision framework that expresses the competing interests in consistent terms and allows a reasonable balancing process. Often the environmental interests and potential ecological impacts are not clearly set out. The developer is always able to show specific economic costs and benefits, with charts, tables, and graphs to make his arguments concrete. The manager must decide issues on the basis of hard, objective facts on one side, and speculative arguments on the other. The environmental concerns are real, and the potential adverse effects are often extremely serious. But a lack of both an adequate biological knowledge

---

243

and a consistent philosophical base hamper the expression of the environmental concerns.

The purpose of this chapter is to explore some of the current limitations on biological knowledge in a broad sense and the alternative philosophical approaches to the problem of extinction of species and environmental management. I hope that a useful guide, providing a context in which the manager and decision-maker can work, will result. It is not possible, however, to provide a detailed guidebook which would tell a government manager exactly what to do in specified cases. The manager must understand that several methods of preservation exist, that they each have limitations, and that many values can be served in the end. The manager's decision will involve a balancing of many factors with the decision determined by the balance.

In order to deal with these controversies, a manager must understand the objectives that he is obliged to seek. First, he must examine his legislative mandate and fully appreciate the values pursued by the competing interests. Second, he must be well advised on the state of the art in biological and ecological research so that he knows what is possible and predictable. Third, he must understand the management techniques that are available to him and the limitations inherent in those techniques. Finally, he must evaluate and apply the management methods most likely to satisfy the objectives and values identified.

I first mention two alternative formulations of the issue of survival. I then catalogue the main objectives currently supported by significant groups, discuss the limitations on management methods, and suggest a strategy for managers.

I. THE ISSUE—SURVIVAL OF SPECIES OR OF THE ENVIRONMENT?

Survival and extinction arguments fall into two major categories. One type of concern deals with the possible extinction of individual identified species, such as the snail darter in the Tellico Dam controversy. At issue was the survival of that one species, and although many other points of economic and environmental interest were argued, that was the single narrow area of decision. The second category of survival arguments is exemplified in the Ehrlichs' analogy likening disappearing species to rivets being popped from an airplane.[2] No particular rivet is crucial for keeping the plane intact, but if enough rivets pop, a crash will certainly result. Likewise, the identity and characteristics of any individual species are not important to the Ehrlichs' central concern. It is the cumulative effect of altering nature by hastening the extinction of species and decreasing biological diversity which leads to the inevitable upset of the global environmental balance.

These two major themes are sometimes used in tandem to argue that the extinction of a particular species will result not only in the loss of that species, but also in extensive disturbance of a specific environment.

The manager must distinguish between these two kinds of arguments, since very different management techniques may be called for depending on which objective is to be pursued. For example, the Bengal tiger may be saved by confinement in zoos if the only purpose is to retain examples of the species. But if we are concerned with the tiger's role in its environment and wish not to upset that role, then establishment of a reserve will be necessary, and the zoo solution will be quite inadequate. We will touch upon these two different approaches many times in later sections.

II. VALUES

Just as society has not made a definitive choice between protecting individual species and preserving natural balances, choices among several more specific values may vary depending on circumstances. These values fall into several categories: tangible benefits to humans; aesthetic enjoyment; biological or ecological roles; and rights and duties. As with the two basic objectives, the manager must decide which of these subsidiary values are to be pursued before he can manage effectively.

Direct Human Benefits

The production of consumable goods is the most obvious direct benefit we gain from the plant and animal kingdoms. Several species have become threatened or endangered because of their interest to man as consumer. Redwoods, most whales, and some fur-bearing animals currently fall into this category, while the passenger pigeon is a sad example of man's overbearing interest in consumption.

A second type of direct benefit is the application of the developed defenses or talents of other species to human problems. The astounding ability of humpback whales to communicate under water over thousands of miles has fascinated researchers and frustrated communications experts as they work to develop comparable systems for human communication under the sea. Penicillin and other antibiotics have led to great improvements in medical practice, and the potential use of other biological and biochemical defenses offers hope for further advances in the treatment of disease.

These values are instrumental to achieving other human goals, e.g., good nutrition and safe medicines lead to health and survival. Benefits must be evaluated in terms of these ultimate goals when placed on the manager's balancing scales.

Aesthetic Enjoyment

Almost any species can provide some aesthetic enjoyment to someone. A biologist may find as much beauty in a primitive microscopic beast as a whale watcher finds in a breaching fin whale. The emotions involved attach to the beauty, the awesome size, the delicacy, and other wonders of the individual creatures, rather than to the endangerment of the species. The possibility of extinction heightens the interest and concern of the human observer because of the fear that the beauty or other wonderful attribute will be lost forever. Again, this value is instrumental to the benefit of human enjoyment and should be weighed as such.

Biological or Ecological Role

Many species play essential roles in the life cycles of other species, such as those which are hosts to parasites or which ensure the pollination of plants. Some play larger roles in the global ecosystem by maintaining the atmospheric balance between oxygen and carbon dioxide. Most species, of course, play some role of importance to other species, with the only difference being a matter of degree. The instrumental value here is more complex, since plants and animals are benefited directly, while humans are benefited only indirectly, through the sustaining of our human environment. This value overlaps the two themes of species preservation and protection of natural balances.

Rights and Duties

Several arguments have been advanced, based on the intrinsic, rather than the instrumental, value of species, assigning rights to species as a class; assigning rights to individual members of selected higher-order species; or assigning rights to the individual members of an endangered species because of the endangerment.

The last of the three approaches underlies the explicit objective of the Endangered Species Act,[3] which protects all members of a listed endangered species and applies penalties for actions against any endangered individual. Unfortunately, this approach is often impossible to enforce, requiring managers to compromise without a guidebook of standards to follow.

The second approach—to protect all members of those species that have some attribute with which we can empathize (e.g., the ability to feel pain)—is the least practical. It requires protecting species, with no concern for environmental balances. If we were to protect each individual animal that can feel pain, for example, the continued existence of *Homo sapiens* in any reasonable form would be impossible. If protection standards as strict as those of the Endangered Species Act (no taking, no harassment, etc.) were applied to all species, life would become impracticable.

This approach has strong underpinnings in human feelings, influencing the manager at a gut response level. Humans easily empathize with individual members of other species when they face death. The closer to man the individual is, either biologically or morphologically, the stronger the empathy will be. Especially in the case of higher-order animals, such as many mammals, man projects himself into the place of the animal, agonizing on its behalf and feeling its pain. Although those feelings of empathy are strongest for individual animals in pain or dying, we broaden the empathy to include the threat of extinction to a species. Our "gut" reaction is to put mankind in the place of the threatened species and to feel the threat of extinction close at hand. These feelings of identification with other species often present the most difficult managerial problems because the real concern is not voiced, but is expressed through a surrogate value.

The first approach, assigning rights to each species as a class, is the most difficult to support philosophically, but is also the most practical. Rights are normally assigned to individuals, not to a class or group. In law, individuals are grouped for purposes of class-action suits, but the amalgamation is for convenience of litigation and represents only the assertion of the same individual rights on behalf of a number of people. However, assignment of survival rights to each species at least permits the death of individual members so long as the species survives. Perhaps the philosophical difficulty can be solved by recognizing that the right of survival is being assigned to the species because of some instrumental value of that species. The instrumental value will survive if the species survives and will not be lost by the death of individuals.

III. Causes of Extinction

The manager, recognizing the objectives and values to be pursued, must bear in mind the major possible causes of the extinction of species.

Habitat Fragmentation and Alteration

Habitat fragmentation/alteration is thought by many to be the most important single cause of extinction today. The draining of America's coastal wetlands, deforestation in Brazil, and general development and expansion in developing countries have the potential for eliminating large numbers of species.

Hunting

The intensely organized efforts to trap passenger pigeons, the global activities of whaling fleets from many countries, and the arrogant killing of Bengal tigers are among the historically important occurrences of extinction through hunting. Although better controlled today, hunting remains a serious problem for many species which have already been reduced in numbers, with poaching still occurring on tigers and Japanese whaling fleets still taking many whales.

Foreign Species in New Habitats

The introduction of foreign species into new habitats, usually done inadvertently, has produced population explosions that have threatened or eliminated other species through competition. For example, black and Norway rats introduced to islands have seriously threatened many bird populations.

Pollution and Other Environmental Changes

Serious adverse effects have resulted from the alteration of global or regional biochemical cycles through acid rain, carbon concentration in the atmosphere, and nutrient flows into the oceans. One of the best documented examples has been the killing of freshwater fish in Adirondack lakes by acid rain.

IV. MANAGEMENT METHODS EVALUATION

The above sections have analyzed two major alternative approaches to survival problems, catalogued several current sets of values, and noted the major causes of extinction, all to be considered by the effective manager. The framework is set, but the resolution of conflict and the search for reasonable answers remain difficult. This section analyzes the major management methods to give a glimpse of the difficulties.

The Endangered Species Act (ESA) Approach

The ESA attempts to protect all individuals of any species of plant or animal that has been listed as threatened or endangered.[4] The manager is not given the flexibility to protect the species as an entity, although that is usually the result in practice.

The limitations of this approach are severe, since the effectiveness of the protection afforded to any species depends entirely on the amount of government resources available. Even in earlier administrations that were more generous with funds, those resources have been very small. Consequently, extreme care must be taken in placing species on the federal endangered list. In the marine area, the listings include several species of whales and sea turtles, some seals, and one species of fish—the short-nosed sturgeon. Even efforts to protect that relatively limited number of species, however, are constantly subject to compromises and choices by the managers.

As an example, after the listing of all of the species of sea turtles, government managers faced a serious problem in the accidental catch of turtles in shrimp nets in the Gulf of Mexico and the South Atlantic. At certain times of the year, turtles are caught incidentally in the shrimp fishery and drown in the nets. It was clearly absurd to close down the entire shrimp fishery, given its economic importance to the nation and to the region. The shrimp fishermen wanted to be responsive, but could not detect the turtles until after the nets were brought aboard to be emptied. The managers decided to develop a cooperative program involving the shrimp industry, the concerned environmental community, and the federal government. The result was a technology development program that produced new gear which allows most turtles to escape from the nets; a training program for the shrimp industry in turtle resuscitation; and continuing programs of information exchange. The method of protection is questionable in terms of the letter of the law, but reflects a practical decision to enhance the survivability of the species, rather than to try to protect each individual.

Other turtle protection efforts involve preserving nesting beaches in order to "headstart" young turtles in a laboratory and release them into the oceans when they are large enough to avoid most predators. Unfortunately, even with the major efforts expended in this case, researchers do not know whether enough is being done to ensure the survival of the species.[5]

The ESA itself contains different approaches to protection. Section 9 prohibits any "taking" (killing, wounding, etc.) of any individual of an endangered species.[6] Section 7 requires that all federal agencies ensure, in consultation with the Departments of the Interior and Commerce, that actions authorized or funded by them do not jeopardize the continued existence of an endangered or threatened species.[7] Under the ESA, a "no jeopardy" opinion under Section 7 by the Departments of Interior and Commerce does not free the other agency from the prohibitions on "taking" contained in Section 9. Consequently, a permittee could be operating in a manner which does not jeopardize the species, but still be prosecuted for injuring an individual animal. The 1982 amendments to the ESA[8] resolve this conflict by eliminating penalties for any takings occurring pursuant to a Section 7 "no jeopardy" opinion. For the first time, Congress has said that the killing of individuals can occur under the ESA so long as the species, as an entity, is not jeopardized. That approach has not, however, been extended to the overall administration of the ESA.

The sea turtle protection program is an example of a strong government and private effort to preserve several closely related species that is reasonably consistent with the objective of avoiding upsets of the natural balance. The values satisfied include aesthetic enjoyment and the rights of species as a class, although probably no other values are encompassed. Turtles are unlikely to be a major food source in the foreseeable future, and they do not appear to play a significant role in the survival of other species. The objective of protecting all individual turtles was certainly not served here by the managers' deliberate decision to allow some level of mortality.

Most importantly, however, the values which are served in protecting these few species of sea turtles can be promoted for only a very small number of species. The National Marine Fisheries Service has made major efforts to protect sea turtles and bowhead whales, with lesser efforts for the Hawaiian monk seal and humpback whale. If the entire effort of one of the two federal agencies charged with the protection of endangered species is consumed with the protection of fewer than ten species, then the ESA approach cannot realistically be used to serve the values laid out above. At best, it can satisfy the values of a small number of people or groups for a very limited number of species. For any of the broader values—preservation of the human race, major enhancement of human living conditions, specific human benefits, or consumable resources—this approach fails.

In practice, the ESA does not protect species generally, since thousands of species are expected to become extinct every decade if present rates continue. The ESA protects only a very few species, chosen for protection by a political process using many different values; nor is the ESA a useful tool for preserving the natural balance. It can be used in a manner that is consistent with that objective, as in the case of the sea turtles, but it cannot protect large sections of the environment from degradation and human-induced change. Its methods can serve only a narrow purpose of protecting a few species. That limitation should be recognized and accepted by the managers. An argument can be made that critical habitat designations under the ESA have a broader protection effect.[9] It is risky, however, to depend on a tool which protects habitats as a by-product of its main purpose, rather than as a stated objective.

Preservation in Zoos

Some species today are preserved only in zoos, where individual specimens are protected from competition, predation, starvation, and other threats. Artificial preservation is of very limited usefulness, however, for fulfilling any of the values set out above. Aesthetic enjoyment can continue, of course, but is severely limited by seeing the animals only in an unnatural environment. The right of the species to exist is maintained, but again, with little impact, since it is not free to exist in a natural way. The preservation of a natural balance, or even a natural state, cannot be achieved at all with this approach.

Preservation of a species in captivity is generally a last-ditch effort to save some individual members of a severely threatened species. It can be used where continued existence in the natural environment is impossible, perhaps because of a major change in a relatively small habitat or because of pressures from many factors which make survival impossible or prohibitively expensive. This approach is itself expensive, of course, and can be used in only a limited number of cases. It also requires a severe compromise in comparison to other methods, since the species is not maintained in its natural state.

Introduction of Species

Serious practical difficulties arise in introducing species into new environments or in reintroducing them into ecosystems in which they once existed. In an old ecosystem, the niche earlier occupied by the species may have been filled by an opportunistic species. In order to thrive in new environments, the introduced species may require different conditions from those it required elsewhere. The ability to predict those needs is very limited with regard to most species. Another factor seriously restricting this approach is its cost, in terms of both the research needed to plan a program and the implementation of programs on a case-by-case basis. Consequently, it is feasible for only a very small number of species.

Introduction of species can enhance only a few of the values that we have been considering. Just as in the case of the ESA, only the values of aesthetic enjoyment and the rights of species as a class can be satisfied, and then only with regard to a very limited number of species. The concerns for natural balance and natural process will not be met, and the overall result could even be negative because of the environmental disturbance.

This approach can be used when the natural habitat of the species has been destroyed or sufficiently altered as to result in irreversible decline of the species. In terms of direct expenditures, it can be less expensive than preservation in captivity, although the risks are higher that the species will not thrive. It may be best to use this approach in conjunction with captive preservation, at least until the threatened species has established itself in the new environment.

Management of Human Interactions with Other Species

This approach is the most effective in situations involving hunting or fishing. Managers can control access to the species involved, limit the number taken or killed, take care that breeding populations and opportunities are adequate, and take other actions that ensure the health and long-term survival of the species. Controls on hunting can be very effective in preventing the extinction of species, despite many examples of failure: the passenger pigeon, Steller's sea cow, etc. Some currently endangered species may fall into the failure column as well, such as the Bengal tiger and the blue whale, although it is too early to know. Controls on fishing usually do not involve species already in danger, but rather those which we want to maximize for commercial advantage. Even in those cases, however, managers have found their ability to achieve stated objectives to be very limited. For example, the Georges Bank groundfish fishery was regulated from 1978 to 1984 with an elaborate series of catch quotas, open and closed seasons, fishing effort limitations, and so on.[10] Yet, the stocks have increased and decreased dramatically over that period in spite of efforts to manage for consistent stock size. The managers have recently returned to a much simplified system, using minimum mesh sizes and spawning ground protection. This system provides adequate conservation of the species in the fishery, but does not provide the precision of management desired.

This method suffers from similar drawbacks to those mentioned above: effectiveness is confined to a relatively small number of species and is expensive in terms of both dollar and personnel resources. This approach can protect the rights of individual species as a class, but not the rights of individual members. It can, when applied to abundant species, enhance living conditions and provide a continuing supply of consumable resources. It can even be used consistently with the objective of preserving natural balances, although it does not actually ensure that result.

Habitat Protection

Several federal and state laws have habitat protection as an objective. The ESA has a specific requirement to designate "critical habitat" for listed species. The Fish and Wildlife Coordination Act[11] provides the authority for the National Marine Fisheries Service and the Fish and

Wildlife Service to comment on the potential adverse impacts on living resources of all federal permit requests. This approach has been used extensively for coastal wetlands, with constant activity involving the two federal agencies just named, the Army Corps of Engineers, and state fish and game agencies. Habitat protection is the only management method which can meet all of the values set out earlier in this chapter.

First, specific benefits to man, such as the discovery of compounds useful in medicine, can be enhanced by allowing species to continue to exist in natural habitats until they can be studied to discern possible benefits.

Second, aesthetic enjoyment is enhanced by habitat protection since the animals or plants are most likely to exhibit their full range of behavior and variation in their natural setting.

Third, the biological or ecological roles of species can follow their natural course since the species continue to live in their own ecosystem.

Fourth, this approach is acceptable to the advocates of rights for species as a class and of rights for individual members, with one important caveat. Those rights are respected only if it is accepted that natural selection is value-free. Species and individuals will sometimes die out in their natural habitat, as they always have, as a natural consequence of interactions with other species and with the physical environment. Finally, habitat protection has the added advantage of being the most effective method of preventing serious threats to currently abundant species. While the manager can protect the long-term viability of some species through direct management of human activities such as fishing, habitat protection may be even more valuable in the long run because of our limited ability to predict the future impact of environmental changes. For example, we know that shrimp in the Gulf of Mexico are dependent on coastal wetlands for spawning and larval growth. We do not know how much of the wetlands can be destroyed before the shrimp are seriously affected. Habitat protection can preserve our options for the future when our knowledge of ecosystem dynamics is likely to have increased considerably.

Returning to the basic dichotomy set out in the beginning of this chapter, habitat protection serves both objectives—protection of individual species and avoidance of interference with natural processes. A central theme in the latter objective is the need for biological diversity on both a local and a global scale. Habitat protection affords the best possibility of increasing, or at least protecting, the diversity of species on the earth. I see diversity, while not an end in itself, as absolutely essential for the survival of the human race, or at least for the survival of a global ecosystem with any similarity to the current one. Some observers believe that we are currently reducing the diversity of species to a dangerously low level, popping enough rivets so that our airplane will soon crash. Others view the global ecosystem as extremely resilient, with little likelihood of human activities reducing diversity rapidly to a point of danger. Regardless of which view is correct, habitat protection is the only approach which can make effective use of limited government and private resources, while trying to satisfy a broad range of objectives.

NOTES

1. *Convention on International Trade in Endangered Species of Wild Fauna and Flora*, 27 UST 1087; TIAS 8249 (Washington D.C., March 3, 1973).
2. Paul R. and Anne Ehrlich, *Extinction* (New York: Random House, 1981).
3. *Endangered Species Act of 1973*, 16 U.S.C. 1531-1543.
4. *Endangered Species Act*, Sections 1538 and 1532(14).
5. Readers interested in further information on this example may contact the author.
6. Ibid., Section 1538.
7. Ibid., Section 1536.
8. "Endangered Species Act Amendments of 1982," U.S. House of Representatives Report No. 97-835, September 17, 1982, pp. 11 and 27.
9. *Endangered Species Act*, Sections 1532(2), 1533, 1534, and 1536.
10. "Atlantic Groundfish Fisheries Management Plan," 50 Code of Federal Regulations 651 (Washington, D.C.: Government Printing Office, 1981).
11. *Fish and Wildlife Coordination Act*, 16 U.S.C. 661-666.

11

Property Rights and Incentives in the Preservation of Species

ROBERT L. CARLTON

I. INTRODUCTION

Few people would argue that we should allow any species to become extinct without first considering very carefully the economic and ecological benefits forgone and the economic and ecological costs incurred. In fact, political decisions to save species have already been made. The Endangered Species Act of 1973, as amended, a sweeping and ambitious piece of legislation, is an example of the commitment of federal, state, and local governments to protect threatened species, even at considerable economic cost. This chapter makes the assumption that it is desirable, for whatever reasons, to maintain biological diversity and that governmental initiatives and efforts to protect species will be undertaken.

Controversies remain, however. Conflicts have arisen because of failures to consider and answer such basic questions as how species should be preserved and who should bear the costs of preservation efforts. Three areas of conflict must be addressed: (1) property rights—can protection of species be achieved while honoring moral, constitutional, and legislative safeguards on property rights? (2) fairness—can means to save species be devised that will fairly distribute the costs of such efforts? and (3) effectiveness—can measures to save species be made more effective by removing disincentives toward compliance?

To date, most efforts designed to protect species have taken the form of legislative prohibitions.[1] The Endangered Species Act, The Eagle Protection Acts, and the Marine Mammal Protection Act, as well as the regulations implementing them, provide examples of such prohibitions by the federal government. State legislation also emphasizes prohibitions against the taking of individual members of a species which is listed as endangered by one or another agency. Although some of the statutes do speak to the protection and preservation of the habitats on which species depend, little habitat protection has actually occurred. But the long-term preservation of species depends above all upon protecting crucial habitats. The failure to safeguard habitats on which endangered species depend, lack of coordination among land-managing agencies and agencies responsible for preservation of species, and general willingness to sacrifice preservation of species for other goals have contributed to the ineffectiveness of the overall effort.

It is necessary to develop and give serious consideration to alternative strategies for the preservation of species. This does not mean forgoing "thou shalt not" approaches, including prohibitions on takings. Such negative approaches, however, can free species from the threat of extinction only if they operate within a general climate in which the majority of those capable of affecting the well-being of threatened species recognize them as having value. In other words, immediate incentives (economic and otherwise) to comply with preservation measures must be provided for those who can affect endangered species. Values that can be realized only in the distant future, if ever, will not be sufficient.

This chapter suggests some new approaches and proposes criteria for choosing among alternative strategies. The three areas of conflict listed above provide an outline of such criteria: any approach to species preservation must respect constitutional and legislative property rights; it must be fair and equitable; and it must be effective. Section II develops the interlocked concerns of property rights and of equity. Section III emphasizes the importance of incentives in developing effective policies. Section IV offers an approach incorporating both incentives and prohibitions into a comprehensive, long-term plan to promote the preservation of species.

II. PROPERTY RIGHTS AND EQUITY

The questions of property rights, fairness, and effectiveness are interrelated. The criterion that any proposed strategy must provide as much protection as possible for private property rights relates not only to equity and fairness, but also to effectiveness. Any policy that is seen as imposing constraints on the rights to hold and enjoy property will be resisted and, consequently, be less effective. Thus, it is effective as well as equitable that the costs of preserving species should be fairly distributed. This means that costs should, to the extent possible, be apportioned among those who will benefit from the policy. As a first, imprecise attempt at stating a workable guideline for preservation policy, one could say that the costs of preservation must be seen as reasonable by those who will bear them. This criterion, however, is too vague and too subjective to be used effectively. In order to make it more objective and precise we will concentrate, first, on concerns about property rights. These concerns lead into the problem of equity. Problems of effectiveness are more easily separable and are discussed in Section III.

Constitutional and legislative protections of property rights are not absolute. State and local governments regulate land use to control public nuisances and harms to citizens and other property resulting from spill-over effects of property use. Further, agencies of federal, state, and local governments can take land for public uses by exercising the right of eminent domain. The distinction between these two broad forms of limitation on the use of property is essential because regulation of the first sort, based on the police power of the state, requires no compensation for interference in the private use of private property, whereas the exercise of eminent domain requires fair compensation for the restriction of rights of property owners.

It is an interesting question whether the protection of endangered species represents a legitimate use of the police power or whether it requires the exercise of eminent domain. If the former, many questions of equity would dissolve: private landowners can be required, without compensation, to refrain from imposing harms or "nuisances" upon the public. If the latter, severe restrictions on private use may be unconstitutional and require compensation to affected landowners.

Unfortunately, judicial decisions have done little to clarify this question, and commentators on judicial precedents have referred to relevant case law as "a crazy-quilt pattern of Supreme Court doctrine."[2] Insofar as discernible patterns have emerged, the following criteria seem to determine the acceptability of compensation-free restrictions based upon the police power: (1) if the policy goal is novel, nontraditional, and perceived as lacking in broad community acceptability, it is less likely to be considered a proper subject of use of the police power; (2) if the regulation obviously benefits a limited portion of the community, it is less likely to be upheld as legitimate exercise of the police power; (3) if the regulation involves acquiring public rights or benefits rather than protecting the public from a discernible harm, it is more likely to be held to require compensation; and (4) if the regulation renders the property in question economically useless or nearly so, it is less likely to be considered a legitimate use of the police power.[3]

While these criteria do not allow a definitive and general conclusion, a case can be made that criterion (2) seems to support the legitimacy of regulation by the police power in the interests of species preservation, as the benefits of species preservation do not obviously serve a minority, whereas criteria (1), (3), and (4) suggest there are circumstances in which regulations imposed by endangered species legislation would not be clear-cut cases of legitimate exercise of the police power. Examining criterion (1), protection of species, except for a few which traditionally have been hunted, fished, or trapped, is a relatively recent concern of the public, and in cases where the species in question is neither well known nor popular, it would be difficult to show broad community support for the goals of the regulation. The protection of species does seem to provide a public good rather than protection of the public against a nuisance, unless an appeal could be made that some rights, either of citizens or of the species themselves, are infringed by the actions of landowners when these actions threaten the survival of a species. Even if one believes in the rights of citizens to have species protected or in the rights of the species themselves, such rights would be moral, not legal, rights. Consequently, legal rights are not infringed by actions threatening species, so endangered species legislation is best seen as creating a public good rather than protecting the public (or other species) from infringements of rights or "harms." Thus, by criterion (3), it would be difficult to justify restrictions protecting species as an exercise of the police power. Criterion (4) is impossible to discuss generally, as the magnitude of the value lost will differ from case to case. Below, cases are given in which the loss is extreme and in these cases, at least, it could be argued that according to criterion (4) compensation is justified.

While this area of law is too complex and the argument made here too sketchy to establish a general case, it seems reasonable to conclude that some of the restrictions imposed by endangered species legislation upon property owners may involve illegitimate "takings" of private property. They may greatly diminish the value of property in the pursuit of a novel, nontraditional public good. At the very least, they involve inequities since individual property owners whose land is inhabited by endangered species may incur significant economic costs which should, more equitably, be distributed throughout the population.

The decision to preserve species is not imposed by biology or economics; it is a political decision. Certainly, it would be very difficult to argue that preservation of species represents an economic enterprise or an economic good, especially for the private sector. Emotion, environmental concerns, and economic concerns combine to create an atmosphere receptive to a political decision to undertake a preservationist policy. The preservation of species redounds to the benefit of all, not just those who undertake preservation efforts or those who have to pay for such efforts individually. The essence of the legislation written pursuant to the political decision to save species is that the preservation of species represents a public good, a good for society as a whole.[4] Unless society as a whole is willing to do without the benefits which accrue from preserving species, the members of society will have to pay for them individually. The question remains, however, whether those costs will be evenly distributed throughout the population.

It is here that the concern for property rights and the concern for equity merge, or at least impinge upon each other. The current approach of largely prohibitive regulations is characteristic of the exercise of a police power designed to protect the public from harms associated with the actions of individuals who threaten the health and well-being of the community. But the preservation of species is a public good, and it is unfair that the burden of the political decision to preserve species should fall disproportionately upon property owners who discover, or are informed, that the presence of endangered species on their lands restricts them from uses of their property not normally considered harmful to the public.

Such cases exist. One example involves the actions undertaken (as necessitated by endangered species legislation) by wood-products companies to protect bald eagles on their lands. One company has more than 900 acres set aside for bald eagle nesting sites and, as additional nesting sites are utilized and located, the number of acres will increase.[5] At current timber values, the unharvestable timber on these acres represents a minimum value of about nine million dollars. Management costs involve at least one man-year per year to ensure protection of the sites. There are also operational costs stemming from the difficulty of logging around the areas that must be left untouched and the timing of tree harvest with the biological patterns of the eagle.

In another example, a forest products company has modeled management of pines which serve as habitat for red-cockaded woodpeckers, comparing their normal rotation schemes with a longer, imperfect rotation which would protect red-cockaded woodpeckers and their habitat.[6] Based on the model, the longer rotation for the protection of the red-cockaded woodpecker would yield $115 less per acre per year than would management which did not protect the species. This decreased return represents a cost to any company that protects the red-cockaded woodpecker by such a rotation.

A number of problems are associated with identifying costs of preserving species. Some costs can be quantified; it is much more difficult to assign numbers to others. For example, some companies have been forced to develop long-range resource management plans on short notice and without the benefit of accurate wildlife or land management data for areas which support species in danger of extinction.[7] If data become available later indicating the management plan must be revised, revision may require coordination with state and federal agencies and the development of a new plan. What does this cost? As another example, in some cases, particularly in the western part of the United States, private and federal lands are intermingled. Very often, federal lands must be crossed in order to reach private land. Before a right-of-way can be granted, federal agencies must be assured that the proposed activities are not likely to affect adversely a federally listed threatened or endangered species. In most instances of probable conflict, alternative courses of action are proposed. Too often, the suggested alternatives are neither prudent nor reasonable, especially when one or more of the agencies involved has little or no knowledge of local conditions or of management possibilities. This can mean delay and in some cases has involved costly law suits.[8]

Government actions can be affected by these kinds of considerations as well. For example, the decision to build the Jersey Jack Road in the NezPerce National Forest was appealed in part on the basis of alleged violations of the Endangered Species Act.[9] Delay to date has affected a $1.2 million appropriation for construction and has adversely affected associated employment and other economic benefits, both nationally and in the region of the NezPerce. The impact of such delays is indicated by the fact that lumber mills in the area employ 500 to 600 loggers and mill workers and provide employment indirectly for another 1,000 to 1,200 people. It is too easy to ignore costs associated with decisions not to take an action or to postpone or shift an action when complications arise in connection with protecting endangered species.

The examples given show that many of the costs associated with the preservation of species are inequitably distributed. The costs of protection measures on private lands have largely been borne by those who have the "misfortune" to own land containing endangered species or on those whose employment depends upon use of such lands. As the examples from the national forests show, local populations of endangered species on public lands can have disproportional local economic effects.

Given that Congress has enacted laws on the premise that the continued existence of species is of paramount concern and value to all of the people of the United States, the question must be asked why the costs are not being borne by all of us. Restrictions on the use of private property may be so serious as to infringe the constitutional right of private owners to the use and enjoyment of their lands. In lesser cases, unfair restrictions may still be imposed upon owners, merely because they, by chance, have populations of endangered species inhabiting their lands. It is unfair to impose burdens upon a minority of property owners, if the benefits are distributed throughout the population. There is a general inequity resulting from the fact that prohibitions differentially affect land use, employment opportunities, and economic growth, depending upon where endangered species are found.

It is possible, then, to state a criterion for judging the fairness of means to protect species: Given that preservation of endangered species provides benefits to all citizens, a method that distributes the costs fairly among the general population is preferable to a method that imposes costs disproportionately on certain property owners, certain areas, or certain geographic or social groups.

III. CONSIDERATIONS OF EFFECTIVENESS

The failure to recognize that the preservation of species is not a free public good, that someone or something will have to bear the cost of such preservation, has adversely affected preservation efforts, in addition to producing inequities. Because our protection system acts through legislative prohibitions, having endangered or threatened species on one's property or under one's management has for the most part been a liability. The kinds of uncompensated costs cited above have often acted as powerful incentives to avoid responsibility for the continued existence of species given legal protection.

Much anecdotal evidence exists of at least local extirpation of legally protected species at the hands of landowners who fear restrictions on their activities. In one case in California a private developer destroyed a population of an endangered plant to ensure that subsequent requests for federal construction grants would not be delayed because of endangered species considerations.[10] In a case in the Southeast, a corporate landowner was informed that an endangered plant was located on its holdings, but when the corporation tried to obtain more specific information on locales, the information was not forthcoming.[11] This made it virtually impossible for the company to set up internal safeguards for the protection of the plants.

This last example points up the need for better cooperation among agencies, and in particular between those agencies concerned with preservation of species and the private sector. Unfortunately, current legislation does not encourage such cooperation to the extent it should. That such cooperation can be very effective in terms of species preservation is shown by the success of the National Marine Fisheries Service in protecting sea turtles from destruction in shrimp nets. The problem arose because the nets in common use until the mid-seventies captured not only shrimp but also turtles. Since the turtles must have atmospheric oxygen and the nets remained below the surface for longer than the turtles could "hold their breath," many turtles captured in the nets were drowned. Although National Marine Fisheries Service enforcement agents could have brought cases against the shrimp fishermen for illegally taking turtles, they decided to work with the fishermen in at least two ways. First, fishermen were encouraged to attempt to resuscitate turtles brought to the surface in their nets in order to return those that might be saved. Second, work was begun cooperatively to develop new nets from which turtles could escape. Such gear was developed and was accepted by the fishermen. This meant the drain on the turtle population due to fishing was effectively halted without major enforcement problems or economic dislocations. However, while this initiative was successful, it was done with no clear legislative authorization.[12]

Indeed, since cooperation with the fishermen precluded prosecuting them for accidental failures of the plan, it undoubtedly violated the stringent antitaking provisions of the Endangered Species Act. An effective program for species preservation must ensure that means for such cooperative efforts are part of any legislative initiative.

In the past, private landowners and public land management agencies have often refused to allow the establishment of endangered species on their lands. They worried that the presence of such species could mean restrictions on land use. Recent legislative actions have relaxed the safeguards for species introduced into new areas, which may alleviate some of these fears.[13]

In general, prohibitions on activities affecting endangered species place landowners who host such species at a disadvantage. They become loathe to report existing populations of such species. Worse, the fear that these populations might be discovered by authorities who would then impose bothersome and perhaps costly restrictions may lead landowners or managers to destroy the organisms before they are discovered. These losses also represent a kind of cost, measured in terms of the diminution of the public good which the preservation of species represents.

If efforts to protect species are to be successful in the long run, they must go beyond prohibitions against the taking of individuals. Concern must be shown for the long-term viability of the species, and this implies increased concern for protection of habitat. Prohibitions do little to create an atmosphere conducive to protecting habitats or to developing methods designed to protect habitat. By such protection, gradual decline of the population can be avoided. A corollary to this point is that an alternative is more attractive if it increases the probability that its implementation will lead not only to a stable situation with respect to population numbers but to increases in its numbers. Thus endangered and threatened species can eventually reach safe and stable population levels, thereby putting their status beyond danger.

These concerns about the effectiveness of endangered species policy suggest another criterion for judging alternative methods of protecting species: Any method that operates, in practice, so as to increase cooperation of the private sector in the identification, protection, propagation, or encouragement of populations of endangered species is preferable to methods that fail to do so.

IV. Conclusion: Incentives and Cooperation

We have established that the preservation of species is a public good. Those who contribute to that effort should be seen as public servants. An atmosphere of cooperation and trust must be created if the effort is to be successful. Such an atmosphere is not promoted by means which depend solely or primarily on prohibitions and legal penalties which simply punish activities adversely affecting populations of endangered species. Such prohibitions and penalties are perceived by those affected as arbitrary and unfair, falling as they do upon individuals who own property or otherwise depend upon the use of property inhabited by endangered species. Under present conditions, the perception of arbitrariness and unfairness compounds the economic disincentives involved in reporting and protecting species, with individuals feeling that their activities are being limited for the sake of benefits to be experienced by others.

A strategy that creates immediate incentives for landowners to report and protect populations of endangered species would be preferable to prohibitions. Any measure has a better chance of success if endangered populations can be identified before they reach the critical stage, i.e., before the loss of single individuals becomes a matter for concern. When the population of a species is perceived to be diminishing but not in danger of immediate extinction, it may be possible to stabilize the situation by setting aside habitat and by taking other steps to provide support to the population. At this stage, prohibitions and the problems they cause are often avoidable. Emphasis can be placed upon measures preventing the decline of populations, rather than on regulations designed to protect each and every individual. Here are several measures that are likely to protect private property rights, to be equitable, and to increase the likelihood of success in species preservation efforts. I have not tried to put them in any order of priority since the most effective approach will vary from one situation to another.

1. Monetary Incentives to Private Owners

These could take any of several forms. They need not be mutually exclusive but rather should be used in the combination most appropriate to a given situation. Among the incentives which could be considered are:

a. Direct payments by that government whose protective measures lead to the incurring of costs. Payments could be made (1) to compensate for damages caused by a protected species in lieu of removing individuals causing damage; (2) to compensate for benefits forgone, such as for the value of unharvested timber in the habitat of a protected species; or (3) to cover the costs of actively managing the environment to meet the requirements of a protected species, such as understory management in red-cockaded woodpecker colony sites. Unfortunately, such payments are highly visible and so would be difficult to sell to legislative bodies concerned with budgetary outlays.

b. Credits against taxes which would otherwise be due to that government whose protective measures lead to incurring costs. Again, this measure might be difficult to sell to budget-conscious legislatures.

c. Allowing costs attributable to protection of species to be treated as expenses for tax purposes. At present such treatment is allowed only if some reasonable expectation of income being generated from such expenditures can be shown. Seldom is this the case. This measure might be more acceptable than the first two discussed since the cost of implementation would be lower.

d. Exemption from all or a portion of state and local taxes for lands on which protected species are located or which provide habitat at some time for such species. Such exemptions would be akin to those employed to protect prime agricultural land from development. This approach would require a high level of commitment from the affected governments since such taxes often form a major part of their incomes.

2. State and Local "Heritage" Programs

Such programs contribute to success in at least three ways: (1) by identifying those elements of our natural heritage that should be saved, thus setting the stage for cooperative efforts; (2) by identifying those elements in immediate danger of being lost unless preventive steps are taken; and (3) by recognizing those corporations, agencies, and individuals whose efforts serve to ensure continued existence of species and habitats. This recognition can be a powerful incentive, as shown by the sums spent on public relations by individuals and corporations to assure being seen in a favorable light.

3. Programs Involving Purchase or Exchange of Rights in Land

In some cases continued private ownership of habitats essential to threatened species may not be feasible. This will mean transfer to a nonprofit organization or public agency of all or some part of the rights inherent in land ownership. Such transfer could be in the form of (1) acquisition of the fee title; (2) acquisition of only those rights necessary to ensure that the lands continue to provide for the needs of protected species; or (3) exchange of the fee title or selected rights to private lands for title or rights of comparable value to public lands which do not provide habitat for the protected species. Acquisition should appeal to those whose lands are not producing industrial outputs; exchange would appeal to those who depend on land to produce outputs for an industry. Leadership in such efforts has been provided by the Nature Conservancy and the Trust for Public Lands. Their programs of acquisition or exchange can provide owners of private lands with timely purchase or exchange so that public and private values are protected; with lower costs for lands or rights acquired; and with various tax advantages.

4. Removal or Easing of Real or Perceived Restrictions on Land

As brought out in Section III, one powerful motive to remove protected species from an area or to refuse to allow their introduction has been the fear the presence of such species would have adverse effects on the ability of landowners or managers to manage lands for desired outputs. Informal conversations with the Office of Endangered Species of the U.S. Fish and Wildlife Service have not revealed any successful prosecutions of private landowners whose land management activities may have adversely affected habitats of protected species, although private suits have been brought.[14] But whether or not fears of prosecution are realized is immaterial—efforts to preserve species have suffered because of them. In some cases restrictions on land management may be needed; however, they should be removed as soon as possible.[15] Provisions for introduction of species at risk into new areas should include few or no restrictions, unless the original populations would otherwise become extinct in the immediate future. Steps in that direction began with the 1982 amendments to the Endangered Species Act.[16] Conversations with public and private land managers indicate that the approach to introduced (or "experimental") populations advocated here would remove much of the disincentive currently felt by these managers.

I do not mean to imply here that prohibitions on takings and restrictions on land use are never to be part of a comprehensive program to save a given species. In emergency situations where the danger of extinction is imminent and where continuing activities contribute to the danger, temporary restrictions and prohibitions will be unavoidable. But such steps will be viewed as less onerous if they are seen as temporary measures required by unusual circumstances. If concurrent positive steps are being taken to improve the overall plight of the species in question, if those efforts are sustained by shared societal resources, and if it is perceived that the restrictions and prohibitions will be lifted as soon as the more general, positive efforts show success, those asked to limit their activities are more likely to be cooperative.

The measures listed above are appropriate for the creation and promotion of a public good such as the protection of species. Any burdens they create are meant to be distributed widely through the society, which is appropriate since the benefits are widespread and long-term. Infringement of property rights of individual owners is minimized and a more positive attitude of cooperation created. As stated in Section I, it is desirable that the costs of preservation be seen as reasonable by those who bear them. We have seen that this implies that the costs of a public good must not fall disproportionately upon those who happen to have populations of endangered species on their property; it also implies that individuals whose economic welfare depends upon the use of land containing endangered species should not be the only ones to suffer from restrictions that threaten their livelihood.

The avoidance of inequitable distributions of costs associated with saving species is not only a demand of fairness and justice, it is essential to the creation of a climate of cooperation which is in turn essential to the effectiveness of the effort. Only when citizens perceive the protection of species as a public good, deserving of a cooperative effort whose costs are fairly distributed throughout the population, will a long-range, effective preservation program be possible. Only then will the underlying problem of maintaining constant support for endangered species be addressed. This approach is preferable to a negative program of prohibitions designed to protect particular endangered individuals. Prohibitions on use of land should be instituted only in unusual, emergency situations, and these situations should be managed with the goal of creating a healthy population as soon as possible so as to remove the need for prohibitions and restrictions on freedoms of citizens.

However, no matter what approaches or means are finally adopted, they should have been selected on the basis of how well they meet the major criteria developed in Sections II and III: those alternatives should be favored which lead to (1) distribution of costs as broadly as possible among the general population; and (2) increased cooperation by the private sector in identification, protection, propagation, and management of endangered species.

NOTES

1. See Steven L. Yaffee, *Prohibitive Policy: Implementing the Federal Endangered Species Act* (Cambridge, Mass.: MIT Press, 1982), esp. chap. 1.
2. Allison Dunham, "Griggs v. Allegheny County in Perspective: Thirty Years of Supreme Court Expropriation," *Supreme Court Review* (1962) quoted in Fred Bosselman et al., *The Taking Issue* (Washington, D.C.: U.S. Government Printing Office, 1973), p. 195.
3. Bosselman, *The Taking Issue*, chap. 10.
4. Yaffee, *Prohibitive Policy*, p. 17.
5. See *Hearings Before the Subcommittee on Fisheries and Wildlife Conservation and the Environment of the Committee on Merchant Marine and Fisheries, House of Representatives* Serial No. 97-32, February 22, March 8, 1982, pp. 261-62.
6. Ibid., pp. 254-60.
7. Ibid., p. 251.
8. For example, see the Gray Rocks case, Nebraska v. Rural Electrification Administration, 12 Envio. Rep. Case (BNA)/1156, 1157 (D. Neb. Oct. 2, 1978) (consolidated with Nebraska v. Ray). It can be argued that Basin Electric might have prevailed, but when faced with interest costs of approximately $100,000-200,000/day, the $7.5 million trust fund was cheaper than a lengthy court case.
9. *Hearings*, pp. 263-64.
10. This involved the occurrence of San Diego Mesa Mint (*Pogogyne abramsii*) on land being developed by a private developer. Informal conversation, June 1980, Office of Science Authority, U.S. Fish and Wildlife Service.
11. This example involves populations of the green pitcher plant (*Sarracenia oreophila*) found on lands owned by a private corporation. Informal conversation, May 1980, International Paper Co.
12. Terry L. Leitzell, this volume.
13. *Endangered Species Act Amendments of 1982*, U.S. House of Representatives Report No. 97-835, September 17, 1982.
14. See note 8, above.
15. See discussion in Section II above.
16. See *Endangered Species Act Amendments*.

Epilogue

BRYAN G. NORTON

To attempt to summarize the discussions and papers that emerged from the meetings of this working group undoubtedly involves overreaching the possible. The members of the group were chosen for the originality of their thinking, not because they adhere to any "party line." Further, the writing of a chapter in such a book as this is usually a contentious enterprise—an important chapter seeks out controversy and defends a novel, not a standard, position. There is a danger, however, that important areas of agreement may be too easily overlooked or ignored. Points of consensus, no matter how hard-won, tend not to be emphasized as much as points remaining in dispute, and claims made in the midst of a controversial essay may carry less authority than when stated as a point of consensus. So, balancing ambition with trepidation, I here attempt to set down major areas of consensus that emerged from our discussions. Stark disagreements will also be reported. Often, it turned out that agreements on policy emerged from important disagreements on scientific or value issues. I see it as especially useful to point out the cases where factual and moral disagreements are neutral with respect to policy options.

All participants agreed that the threat of widespread species extinctions is very real. And while current rates and future projections are subject to debate, there is no question that conditions favorable to mass extinctions now exist and are becoming more prevalent. It was also agreed that far too little is being done to avoid extinctions, given the magnitude of the potential effects.

I. MULTIPLE PROBLEMS INTERTWINED

The goals of preservation are stated in varied ways. Some express concern for protecting gene pools, others for saving extant species, and yet others for reversing trends whereby extinction rates are increasing. Since eco-systems undisturbed by intense human use provide a wide range of human services and amenities, it is not always clear whether preserving species is an end in itself or whether it is a means to protect ecosystems in less disturbed states. Members of the group showed concern for all of these goals, but emphases on them varied.

Therefore, the problem of endangered species is best viewed as a cluster of related problems, not as a single one. (1) What is to be done when the population of a particular species approaches a minimal threshold below which its continued existence is problematic? (2) How are relatively stable healthy populations to be kept that way—how does one avoid situations where healthy breeding populations decline toward a minimum threshold? (3) What can be done to halt or at least retard the seemingly inevitable acceleration of the downward trend in biological diversity that has been projected to occur over the next 25-30 years? (4) How can the increasing tendency of human beings to alter natural systems for intense use be slowed?

The working group attempted to consider all four questions. These problems are interdependent, and yet they must be addressed in quite different ways. To the extent that all are rooted in socioeconomic trends—economic growth based upon high degrees of waste, increasing con-sumption of resources in developed countries, and population growth around the world—they are in one sense identical. If these socioeconomic trends are irreversible, problems (2)-(4) will inevitably be addressed in the form of (1). The defense of natural diversity will be no more than a series of continuing and eventually hopeless rear-guard actions.

Problem (1) encourages an individualized species-by-species approach. The movement from (2)-(4) reflects a progression toward a more holistic conception. Expressions of concern for loss of species often indirectly represent a concern over the progressive alteration by humans of more and more natural habitats and over the further degradation of previously altered systems. The extinction of species is taken as an indicator that human activities are damaging the overall functioning of ecosystems. Is concern for ecosystems a mere byproduct of concern for species? Or is the preservation of species valuable mainly because it contributes to the protection of ecosystems? Indeed, one might wonder whether the value of species can be separated from the value of ecosystems. These questions were discussed at length, but no consensus was reached. They represent a fruitful area for research in environmental values. We will see, in Section IV, that some important management recommendations can be made without resolving species/ecosystem issues.

II. Scientific Background

A. Rates and Trends

The working group took no position on rates of extinction, except to note that the rates currently quoted are based upon speculative data. All believe that rates even at the bottom of the range usually cited would be extraordinary in evolutionary history. There have, of course, been major "extinction events" in the past, marking the end of geologic and climatological eras. While these events may have approached currently expected species losses in total numbers, they were spread over much longer periods of time, measured in millennia. We can say that the coming wave of extinctions is unprecedented in the last 60 million years, especially in abruptness but probably also in total number of species that will have been lost.

B. Causes

Extinction events occurred before human beings existed and some current endangerments are independent of pressures from humans. The very large increase in the rate of extinctions, when combined with observations of widespread alteration of other species' habitat by humans, however, provides strong evidence that a large majority of current endangerments are due to human causes. No consensus emerged concerning what, if anything, should be done about species endangered by nonhuman factors. If, as a minority believed, the major concern is to preserve as many extant species as possible, the cause of endangerment should be irrelevant. But most participants felt that the growing rate of extinctions should be the major concern, and on this view it is reasonable to concentrate efforts on the majority of current endangerments due to human causes. On this conception, human activities are responsible for increasing the extinction rate and steps should be taken to redress this tendency, but it does not follow that steps need be taken to combat nonanthropogenic threats to species. Such losses of species can be seen as "natural," with the possibility of being offset by naturally occurring speciation events.

Humans cause extinctions by habitat fragmentation, degradation, and destruction, by exploitation, by pollution, and by (intentional or unintentional) introduction of exotic species. Habitat fragmentation and destruction and exploitation appear to be the most important of these. Human alterations of the environment and changes in human technology occur very rapidly, leaving little time for other species to adapt. Behavioral adaptations, such as restriction of range, switching to nocturnal habits, etc., may occur in response to these rapid changes, but the period of time over which the changes occur does not usually permit adaptations to become fixed in a population. Consequently, human-caused extinctions may be more similar to natural geologic and climatological disasters than to adaptational pressures placed upon a given species by competing or predatory species. That is to say, human pressure on the environment is not similar, evolutionarily, to normal selection pressures exerted by other species. This line of reasoning undermines apologies for human-caused extinctions as "natural" and as a part of the evolutionary process not requiring special concern. If the rapid changes brought about by technologically augmented human activities preclude genetic responses by other species, and if they are of such magnitude as to cause events resembling the mass extinctions of past periods of the earth's history, then humans must, if they are to avoid radical changes in life patterns on earth, at least take steps to minimize the effects of their activities.

C. Effects

Human activities do not affect all species equally. Human alterations of the environment tend to have their greatest effects on species that occupy upper levels of food pyramids, have large individual size, have relatively small populations, reproduce relatively slowly, and live in specialized habitats. These species are less able to respond to large losses in numbers and can more easily be driven to dangerously low population levels.

Human-caused extinctions, then, are not only discontinuous in type and rate compared with nonhuman-caused extinctions, they also cause selective skewing of evolutionary trends. Specialized species are more likely to have highly evolved relationships (symbiotic, predatory, competitive, etc.) with other species in their environment, and the loss of such species may cause further disruptions and can result in a cascading effect through ecosystems. Because high degrees of specialization to biotic and abiotic habitats preclude interchangeability of specialized species, when these organisms are lost their niches are usually filled by less specialized, opportunistic species; in this sense, systems become progressively simplified. Consequently, the effects of extinctions should be measured in terms of the progressive simplification of systems as well as in numbers of species lost from all systems.

Since so little is known of how species interact, it is usually not known which ones are essential for the continued functioning of a given system. Considerable redundancy in systems guards them from early breakdown, but the loss of specialized species encourages replacement by weeds and pests. Selective extinction, then, may create a world with fewer highly adapted, specialist species that are useful and pleasant to humans and more members of a few opportunistic species. While such species may

271

be familiar (for example, kudzu as ground cover, water hyacinths) their population patterns can thwart human goals.

We know from the fossil record that the effects of major disturbances filter through systems very slowly, and the full effects of extinctions may not be known for centuries. The great extinctions of the past were followed by long periods of reduced diversity and by a long, slow rebuilding process, as the remaining stock of unspecialized species gradually adapted to special niches, and natural selection led once again to a more diverse world. Although such processes require millennia, the recognition of this time-scale should not deter us from doing our best with current information and intellectual tools.

III. Values and Policy

Reasons and arguments for saving species are most conveniently categorized according to their value premises. Empirical premises play some role in all arguments for saving species, but a policy recommendation can only follow if some value or goal is pursued, and it is the value premise that gives various arguments their distinctive character. There are no value-free scientific arguments for or against species preservation— or for any other decisions.

A. Homocentrism vs. Nonhomocentrism

It has been customary to distinguish the homocentric reasons humans might have for saving species from the reasons that can be derived from a belief in the intrinsic value of nonhuman species. But the working group concluded that this much-discussed dichotomy is less important than is usually thought. No member of the group felt that value exists wholly independently of human valuation. Controversies remained concerning the types of dependencies existing between values of other species and human values, but arguments concerning whether particular values attributed to other species should be called "intrinsic" or not proved to be largely matters of terminology. For example, some participants believe that other species contribute to values quite independently of what humans in fact value, but that the value inevitably presupposes human valuing. Another believes that concern for other species is an expression of a natural human sentiment of altruism. In these cases, the value of other species is independent of what individual humans want or desire, but it still occurs on a human scale. It seems more important to understand the ways in which nonhuman species contribute to human values than to decide whether any particular value is best labeled as "intrinsic."

There was agreement that there is an important difference between reasons for saving species (and corresponding values of them) which derive from fulfillments of desires or needs actually felt and expressed by humans, on the one hand, and those which arise independently of such human preferences, on the other. Because participants were convinced that there are some reasons of the latter sort, there was no agreement that species preservation should be justified purely on the basis of the ways in which nonhuman species serve human, stated interests. It is possible, then, to categorize reasons for saving other species as human, utilitarian reasons—those values stemming from recognized, individual human interests—and nonutilitarian reasons—those which transcend such interests. There was little support for attributing value to other species entirely independent of human values, although some preferred to call values independent of utilitarian ones "intrinsic."

B. Human Utilitarian Arguments

The working group agreed that, regardless of one's position on homocentrism, there are utilitarian reasons of varying degrees of persuasiveness for humans to preserve other species. The most common of these appeal to commercial (and other) uses. Since many species provide commercially valuable products, all species should be saved, the argument goes, because we can never know which of those not examined will prove useful. The group did not find this argument persuasive, because the probability that any given species will be commercially useful is relatively low. Further, until a particular use is proposed, there is no way to place a quantified commercial value on potential benefits, nor to decide whether a given species is uniquely able to provide them. As important as considerations of potential use may be, the group feared arguments based on this premise are unlikely to be decisive in policy debates because the relative uncertainty of discovering such uses compares unfavorably with the comparatively "hard" economic data often cited in favor of development projects.

Other utilitarian arguments are more convincing. Humans derive many important services from ecosystems. Ecosystems purify water by providing tertiary treatment of human wastes, stabilize the hydrology of an area by retaining ground water, and sustain the quality of the atmosphere by maintaining the proper mix of oxygen through photosynthesis by plants. Insects in wild ecosystems fly into managed areas and pollinate many crop species, and predators persist in natural ecosystems ready to attack in the event of an outbreak of pests.

In addition, ecosystems are a major source of aesthetic pleasure. Since ecosystems are composed of species interacting with their physical environment, species are the ultimate providers of these benefits. Even though such services are usually noncommercial and not easily priced, it is, however, a certainty that the cost of providing them artificially would be very high in terms of both money and energy. Thus, the several values humans derive from ecosystems are great enough to justify even expensive preservational efforts.

The group agreed that the most persuasive of utilitarian arguments for species preservation derives from fear of ecosystem collapse. This argument gains strength with each extinction since some minimal set of species is necessary to sustain human life. As each species is eliminated, the redundancy within ecosystems is reduced and the fear increases that the next loss will cause a serious disruption in ecosystem services. Since species depend upon each other, each additional loss makes further losses more likely. Because humans derive so many benefits from the functioning of ecosystems, this argument was given great importance by the group. The kind of simplified biotic world that scientists have recognized as characteristic of periods of mass extinctions would not be a pleasant one for humans to inhabit; and the processes that reestablish diversity would require millennia.

The view has been put forward that these important values can be stated in quantitative as well as qualitative terms, and controversy exists over the usefulness of benefit-cost analysis as a method of deciding which species preservation efforts are socially beneficial. The advantages of trying to quantify the values of species are considerable, since quantification brings the reasons for preservation into the mainstream of the decision process. And if the categories of value are construed broadly enough to include the full array of benefits derived from species, the preservation option often receives extremely high marks. It is important to recognize that rigorous economic analyses often do not favor developmental activities over preservation.

However, as broader categories of value are comprehended in the analysis, it becomes more difficult to advance noncontroversial dollar-value assignments. Commercial uses of species fare pretty well in this context, but aesthetic values and various other services of ecosystems are far more difficult to quantify. Even if one could assign a replacement cost to the service in question, scientists lack precise knowledge of the comparative contributions of species to the provision of services. Therefore, it is unclear how to apportion benefits to the individual species that compose a system. Similarly, some economists cite the arbitrariness of quantifying the risks associated with ecosystem collapse. As with other low probability events with great negative consequences should they occur, qualitative goals such as reduction of known risks may be more relevant than attempts at precise quantification of all risks.

Deciding the proper time frame over which to compute costs and benefits may be the most serious difficulty with quantitative methods. Since species, once lost, cannot be re-created, a decision to allow a species to become extinct is irreversible. Any values accruing to the human race through indefinite time would be lost. But quantitative analysis must assume some limited time frame and cannot, therefore, represent the fullest, long-term value of a species. In general, the value of a species appears to increase as the time frame over which benefits are computed lengthens. Decisions to preserve a species become more defensible as one places greater weight on the obligations of present generations to protect the future.

The working group could reach no consensus on the overall usefulness of benefit-cost analyses in decisions concerning endangered species. Nobody advocated their use as a substitute for public debate and decision-making in political terms. Few objections were raised to their use as a means to clarify the economic aspect of such debates. Insofar as a type of benefit is, in principle, priceable, proposed cost figures can serve as hypotheses open to debate, criticism, and falsification. But, given that detailed knowledge of ecosystem functioning is often lacking and that noncommercial values are not easily assigned dollar figures, benefit estimates are susceptible to self-interested manipulation. Therefore, benefit-cost analyses should be used with caution.

C. Nonutilitarian Arguments

Many preservationists, while believing that other species are useful to humans in important ways, also believe that species have value in their own right. The claim that nonhuman species have intrinsic value is, however, difficult to interpret and defend, especially if this claim is taken in its strongest form as implying that nonhuman species have value entirely independent of human assessment or recognition of it.

One attempt at clarifying independent values, that of interpreting claims of intrinsic value in terms of the rights of nonhuman species, was rejected as not helpful. Traditionally, rights have been attributed to human individuals. Extensions of this concept to nonhumans therefore fails to provide appropriate support for protecting *species*; concern for rights of nonhumans cannot explain the preference preservationists give to the last remaining members of an endangered species (over the individually threatened members of a thriving species). Attributions of rights to other species are, at best, useful as an expression of moral concern and commitment or as a forensic device. But they provide no theoretically defensible basis for species preservation.

But it is unnecessary to posit values for other species entirely independent of human value systems in order to recognize nonutilitarian values of other species. It was not agreed that strongly independent intrinsic value assigned to nonhuman species was completely indefensible. The group felt, however, that the clearest theoretical arguments assigning nonutilitarian values to nonhumans always take account of a human role in value attribution. Human altruistic sentiments toward other species may give rise to motives not reducible to the recognized preferences which are the basis of utilitarian justifications. Further, human learning experiences deriving from contact with other species may alter rather than fulfill human preferences. Such experiences, which elevate human consciousness, have value whether they fulfill some felt preferences or not.

The range of human motives should be seen as much broader than the recognized, often consumption-oriented preferences that form the currency of utilitarian justifications. The role of experiences of nature and other species in forming human ideals and aspirations should be recognized and should force us to go beyond the limitations of the strictly utilitarian approach. If we were to do this, we would find that it is possible to cite a wide variety of values served by nonhuman species, without insisting that other species have value entirely independent of the process of human valuing.

Having minimized the distinction between homocentric and purely nonhomocentric motives for protecting species, the group agreed that the strongest, most comprehensive rationale for saving species would appeal to a very wide range of human motives and values. These motives derive, first, from the material and aesthetic uses to which humans put other species; second, from services derived from ecosystems (which are, of course, composed of species in a particular environment); and third, from concern about ecosystem breakdowns that could threaten continued human life. The group also agreed that humans can and should recognize higher, less immediate, and less material values in other species. Immediate needs and desires provide no sufficient basis for human values in general, and there is no reason to rest the case for the preservation of species solely on them. Humans choose ideals and goals not just as individuals but as societies, cultures, and as a species; the rational, spiritual, and emotional bases for these ideals do not represent a mere aggregation of felt needs and desires. Indeed, those ideals create and shape, to a considerable degree, the needs and desires felt.

Thus, as one looks beyond immediate needs, ideals such as living in harmony with nature and keeping the world a healthy place for our children and grandchildren to live in provide a powerful collection of reasons for preserving other species. The more one sees humanity as having an ongoing, indefinite, constantly changing, and evolving history, the more likely one is to give less weight to immediate monetary needs and desires. In this larger perspective where the values of the future must be shaped as well as fulfilled, immediate commercial advantages are balanced by higher human values. The range of values cited, then, encompasses the felt, immediate needs of individual humans, but also goes beyond them to recognize that other species should be protected as a matter of human "good behavior." These precepts of good behavior are not exhausted by concern for the well-being of other living humans. They also include concern for the future of the human race and concern for the rational, spiritual, and emotional ideals that shape that future. The group agreed that, using this expanded conception of value, we can find powerful and convincing reasons to preserve nonhuman species.

IV. Scientific Limitations and Management Recommendations

A. Scientific Limitations

It was overwhelmingly agreed that the lack of scientific knowledge regarding ecosystem functioning is deplorable. Since there are few noncontroversial theories describing ecosystem functioning in general, management efforts usually depend on gathering specific information regarding particular species and ecosystems. In several areas more background knowledge would be especially helpful.

Levels of redundancy—the numbers and types of species that are expendable without precipitating ecosystem collapse—are not well understood. Nor is it known whether most systems have comparable levels of redundancy or whether these levels vary greatly from system to system. Further, it is not known if some, many, or most ecosystems are governed by one or a few "keystone" species—species that determine a system's character and functioning, without necessarily being dominant numerically or in terms of total biomass. In general, information on dependency relationships is badly needed for sound management efforts.

As was noted above, some controversy arises over whether concern should be directed toward the comparative rates of extinction and speciation or toward the protection of the particular set of species that is currently extant. If the former, it would be useful to know what variables encourage speciation. Many of the same pressures that push a species toward extinction also make speciation more likely. Habitat fragmentation and altered adaptational pressures (two results of changing land use patterns) should promote speciation. Normally, human alterations of habitat do not have this effect because the changes occur too quickly for other species to adapt and because the isolated population fragments are subject to continued but changing patterns of human interference so that adaptations usually do not flourish.

Overwhelming these gaps in general knowledge, however, is the lack of specific knowledge concerning the life histories, habits, and relationships of many individual threatened species. In general, when a species is identified as requiring protection, the process often begins with little knowledge of its needs. While this problem is pervasive, it is especially acute in the tropics, the site of most projected species losses. A majority of tropical species have neither been identified nor named. Few have been studied in any detail. Many species will be lost without our having even known of their existence.

But the scientific situation is not as desperate as these areas of ignorance might suggest. Falling between the extreme of encompassing generalizations about ecosystem functioning and the lack of detailed information lies a vast store of middle-level knowledge that can and will prove valuable to managers. These are generalizations of intermediate scope about the habits and characteristics of species. This information is valuable in indicating which species are most prone to extinction and which are protected by their individual or populational characteristics.

B. Habitats or Species

This epilogue began by isolating four separate but related problems, ranging from what to do when the population of a species has declined to the point where its continued existence is in jeopardy to a general concern for the continuing human alteration and simplification of natural ecosystems. Not surprisingly, problems as diverse as these demand more than one management strategy. Two general approaches—the species-by-species strategy and the ecosystem protection strategy—can be identified.

The species-by-species approach, represented by the various Endangered Species Acts that form the cornerstone of the U.S. federal government's most organized and best-publicized efforts at species preservation, begins with an inventory of species believed to be threatened. Preliminary lists of such species are winnowed and the final choice of species requiring individual attention is announced. A set of management procedures, perhaps including a designation of critical habitat, is proposed and, ideally, these procedures are put in place to protect the species. Habitat protection may be an important part of the management plan, but it need not be.

The ecosystem protection approach, represented by the Nature Conservancy and other private ecosystem protection groups, typically inventories representative ecosystem types for a geographical area and attempts to set aside from human use one or more examples of each type. The existence of endangered species is not irrelevant to this process, as knowledge that an endangered species exists in one ecosystem may be a reason to give it priority in choosing areas to protect. But concern for ecosystems is primary.

The species-by-species strategy is likely to be appropriate when a crisis situation is recognized with respect to a particular species. The group concluded, however, that the ecosystem protection approach has considerable advantages in all noncrisis situations. In short, concern for individual species should be limited to crisis situations and, even then, management initiatives should be designed to rebuild the population of the threatened species so that wise management of its habitat can replace direct intervention in the life cycle of the species. Several reasons support this conclusion.

The first reason favoring habitat protection is the lack of knowledge mentioned above. Species are most likely to fulfill their essential needs in undisturbed habitats. Managers, having imperfect knowledge about how ecosystems function, often do not sufficiently understand the needs of species or how to fulfill them to preserve the species. Thus, while managers may have to intervene when habitats have been too badly disturbed to provide a species in crisis with the support it needs, long-term success depends upon reintegrating the species into a functioning habitat as quickly as possible.

Second, while no full agreement was gained on whether ecosystems have intrinsic value, it was agreed that ecosystems, as well as species, have instrumental value. Ecosystem protection saves both species and ecosystems. It is therefore immaterial whether the value of ecosystems is regarded as intrinsic or as instrumental. This is one point at which an interesting philosophical debate does not affect managerial choices.

Third, protection of ecosystems has positive side effects for all species within it. Often it is much easier to obtain public support for the protection of large, aesthetically pleasing species. By protecting a habitat sufficiently large to accommodate such species, rather than placing them in zoos, special preserves, or botanical gardens, many other species will be preserved in the process. Further, an extensive system of protected ecosystems can avoid population declines that might cause crisis situations.

A fourth advantage of the ecosystem approach is that it focuses upon the long-range problem. Last-ditch efforts to save species in crisis situations are expensive and may be only temporarily successful if a suitable natural habitat for the species is not also protected. Or, if the effort is successful only in establishing populations in zoos or highly managed reserves, only some of the values derived from saving the species will be realized. Efforts expended on ecosystem protection, however, offer a chance for the species to perpetuate itself in its natural habitat with some hope of permanency.

Two qualifications to these advantages are worth noting here. First, with very large-bodied, top-of-the-food-chain predators (mammals and birds) where successful breeding populations require very large ranges, intense management at the edges of preserves plus efforts at captive propagation may prove necessary even with the ecosystem protection approach. It may be too expensive and too socially disruptive to provide a completely natural habitat for tigers, while a managed habitat and captive propagation may work acceptably. Since such species' ranges are so much larger than those required by other species, some concessions may be justified. It is important that the concessions not be carried too far, however, since it is the largeness of the range of higher-level predators that supports the decision to tailor the size of habitats to their needs.

Second, protection of a broad gene pool for species important in food production and development of pharmaceuticals may require special preservation efforts. Protection of a few natural habitats may not be sufficient to ensure that we have a sufficiently broad spectrum of genetic characteristics for these purposes. The possibility of creating property rights in such gene pools might encourage the private sector to undertake some of this work. Such property rights may, however, be problematic in other ways. More study would be useful here.

Even with these qualifications, the group strongly supported habitat protection as the most effective approach to the endangered species problem, for both practical and theoretical reasons. It can make use of available scientific knowledge while fulfilling the broadest possible range of human values, thereby avoiding disagreements over the exact nature of the problem and over value rankings.

These generalizations, then, imply that the current U.S. policy on endangered species is skewed, in an unfortunate manner, toward concern with specific crisis situations. While such efforts form an essential part of a comprehensive protection policy, more attention should be paid to the broader problems of wholesale habitat alteration and worldwide loss of biological diversity. The current U.S. government policy of concentrating on particular endangered species cannot be transferred to tropical areas where more than half of the species are not even named, let alone identified as endangered. A general policy shift from crisis management to a comprehensive effort at ecosystem protection worldwide is essential. It may be necessary to expend less effort in saving some domestic species, in order to try to address the more pressing international problem. These changes will no doubt involve increased expenditures, as well as shifts in usage of currently committed resources. Only in this way will the full range of policy problems lumped under the heading "the problem of endangered species" be effectively addressed.

C. Priorities: The Triage Problem

An important charge to this group was to address the question of priorities. It was hoped that some generalizations would emerge concerning which species could and/or should be given highest priority when financial and managerial resources are insufficient to allow major efforts to save each and every threatened or endangered species. Little headway was made in addressing this question in this form.

Many have thought that priority rankings should be based upon whether a species is "high" or "low" on the phylogenetic scale—assuming that the values humans derive from other species vary according to this factor. But this is not the case. Concern to protect sources of edible biomass and ecosystem services would suggest priorities be placed on plant and animal species low on the phylogenetic scale, while concern for aesthetic values would suggest special treatment for mammals and other vertebrates. We come to an impasse unless agreement is reached on whether productivity and material values or aesthetic values are more important. No societal consensus on these value issues is likely to emerge, and rankings based upon the phylogenetic scale are therefore unlikely to achieve widespread support.

More agreement can be obtained by basing priorities on the distinction between specialist and nonspecialist species, but the recommendations derived from this approach are not generally useful. It can be said that nonspecialist, opportunist species are usually less valuable as resources for human use. They are often weedy, pestiferous, and harmful and annoying to humans. So aesthetic and productivity considerations converge upon specialists. But little follows, because generalists are virtually never endangered. They are, by definition, capable of occupying varied habitats, often including those altered by humans. Such species are unlikely to require human assistance, so it hardly relieves the burden of species preservationists to know that they are less valuable.

These negative results are no reason for despair, however. As is often the case in attempts to answer recalcitrant questions, the result is a transformation of the question asked. Having made no progress in species-by-species importance rankings, the group found itself discussing answers to a different but related question: given limited resources in money and effort, how should managers expend those resources? This question does not assume a species-by-species ranking and opens the field to a wider range of answers. And, once the question was framed in this manner, a strong consensus emerged: in general, our resources should be directed to protecting habitat rather than to managing individual species. This recommendation might effectively guide resource expenditures, while avoiding the seemingly impossible task of species-by-species rankings. The reasons have already been given above for favoring this strategy whenever possible.

This, of course, does not suggest that heroic efforts to save particular species are never justified. Some species are known to be useful to humans; others are of great aesthetic or cultural interest. If populations and habitats of such "special" species are already so depleted that mere protection of remaining habitats would fail to protect them, intense management efforts such as captive propagation may be justified. Likewise, additional measures such as seed banks and preservation of animal germplasm have an important role both as insurance against unexpected extinctions in the wild and as a means to preserve a broad genetic base as wild populations decline. These methods are especially important for crop species, their close relatives, and other species with significant utilitarian values, but they do not replace habitat protection and preservation in the wild. Species preserved in the wild continue to interact with other species and to adapt to changing conditions as species preserved by artificial means cannot.

Some observers might suggest that new technologies permitting genetic manipulation may make *in situ* preservation less important. Nothing could be further from the truth. Species can only be created anew if blueprints exist to guide the process. Also, the possibility of genetic manipulation makes existing genes and genetic combinations more valuable—the more "building blocks" available, the greater the number of combinations and recombinations possible. Species in the wild are repositories for existing combinations and they also create new ones, thereby enriching the promise of manipulative technologies.

These various artificial means of preservation should not be the main thrust of a national policy of species protection. Nor should they be undertaken as a matter of course just because a species is nearing the brink of extinction. Unless there is some positive reason to believe a species has utilitarian, aesthetic, or cultural value to humans, or that it has special inherent characteristics such as intelligence, then the present line of reasoning suggests that efforts to list and save it may be ill-placed. If the species has no documented value, then its value is presumably comparable to that of unidentified species. We have convincing arguments that more species are preserved by saving large and representative habitats than by expending limited resources in many crisis situations. Furthermore, efforts to protect ecosystems will pay off in solutions to the more basic problems. They will, in the long run, help us avoid bringing more species to the brink of extinction and protect biological diversity. Ecosystem and habitat protection, then, protects more human values, serves wider human goals, and, ultimately, saves more species than do expensive efforts to protect particular species.

There is no doubt that in attempting to preserve biological diversity we will impose costs on individuals and on society. Members of the group felt it reasonable to assume that, in developed nations such as the United States, efforts to preserve species are affordable and the benefits far outweigh the costs. Setting priorities can be seen, then, as choosing how to spend resources that are available for species protection so as to achieve the greatest possible benefits per dollar of costs.

The situation is far more complex in less developed countries where basic human needs are unmet. This is an important point, as many threatened species exist in the less developed nations of the tropics. In such countries, the costs incurred might well limit the fulfillment of basic human needs. The group agreed that the preservation of biological diversity is likely to contribute to the long-term health of any national economy. In the short term, however, in a poor and crowded country, it may be difficult or even an impossible luxury to take the long view. If so, more affluent nations must play a role in habitat protection in less developed nations. The problems of sovereignty and cooperation will be very complex, but efforts to overcome them seem essential.

Notes On Contributors

J. BAIRD CALLICOTT is Professor of Philosophy and Coordinator of the Letters and Sciences Program in Environmental Studies at the University of Wisconsin-Stevens Point and is on the editorial board of *Environmental Ethics*. Professor Callicott taught one of the first courses in the United States in environmental ethics and is a scholar of the work of Aldo Leopold. He is coauthor of *Clothed in Fur and Other Tales: An Introduction to an Ojibwa World View* and has published numerous articles in philosophical and conservation journals.

ROBERT L. CARLTON has served as Manager of the Wildlife and Non-Timber Resources Programs of the National Forest Products Association and as a Wildlife Specialist for the Cooperative Extension Service of the University of Georgia. He has also taught biology (at Concord College) and been a fisheries biologist for the U.S. Fish and Wildlife Service. He has written on various aspects of wildlife management for the Cooperative Extension Service and contributed to such publications as the Proceedings of the North American Wildlife and Natural Resources Conference.

STEPHEN R. KELLERT, Associate Professor at the Yale University School of Forestry and Environmental Studies, is Director of the Richard King Mellon Nonprofit Conservation Organization Program. He is Associate Editor of the *Journal of Leisure Sciences* and has received a Special Conservation Award from the National Wildlife Federation. He has written widely on attitudes toward wildlife.

TERRY L. LEITZELL is currently an attorney with Bogle and Gates in Washington, D.C., specializing in ocean resource law. He was Assistant Administrator of the National Oceanic and Atmospheric Administration for Fisheries from 1977 to 1981, with responsibility for administration of the Endangered Species Act. From 1970 to 1977, he was a chief negotiator for the United States in the United Nations Law of the Sea negotiations, with primary responsibility for the treaty sections on protection of the marine environment and marine scientific research.

THOMAS E. LOVEJOY is the Vice President for Science of the World Wildlife Fund, Chairman of Wildlife Preservation Trust International, and a member of two commissions of the International Union for the Conservation of Nature. He is coprincipal investigator of "The Minimal Critical Size of Ecosystems Project" jointly administered by the World Wildlife Fund and Brazil's National Institute for Amazonian Research. This large study is designed to quantify, through biogeographical analysis, the optimal size of natural refuges. Dr. Lovejoy is the author of more than eighty publications.

BRYAN G. NORTON was a Research Associate at the Center for Philosophy and Public Policy at the University of Maryland, while on leave from New College of the University of South Florida where he is Professor of Philosophy. He is the author of *Linguistic Frameworks and Ontology* and articles on various philosophical subjects, including a series of three essays on the foundations of environmental ethics, published in *Environmental Ethics*.

ALAN RANDALL is a Professor of Agricultural Economics at the University of Kentucky. He is the author of a leading textbook on natural resource and environmental economics and of numerous articles in economics journals. Currently, he is leading a major research project to estimate the benefits of the national environmental protection effort.

DONALD H. REGAN is Professor of Law and Professor of Philosophy at the University of Michigan. His book on the structure of consequentialist moral theories, *Utilitarianism and Co-operation*, won the Franklin J. Matchette Prize of the American Philosophical Association. Professor Regan has been a Senior Fellow of the National Endowment for the Humanities and a Visiting Fellow at All Souls College, Oxford.

LAWRENCE B. SLOBODKIN, Professor of Ecology and Evolution at the State University of New York at Stony Brook, has contributed for more than thirty years to the development of evolutionary and ecological theory, including work on predation theory, problems of perturbation, and ecological management. He is currently the representative of the New York State Attorney General to the Hudson River Foundation.

ELLIOTT SOBER is an Associate Professor of Philosophy at the University of Wisconsin-Madison. He spent a year at the Museum of Comparative Zoology, Harvard University, supported by a John Simon Guggenheim Foundation Fellowship, and has written articles on philosophy of science, metaphysics, and epistemology. He has also published an anthology, *Conceptual Issues of Evolutionary Biology*, as well as his own book, *The Nature of Selection*, both with MIT Press.

GEERAT J. VERMEIJ is Professor of Zoology at the University of Maryland and a Research Associate at the Smithsonian Institution. He has worked on the ecology, functional analysis, and evolutionary history of shelled marine animals. Professor Vermeij is the author of *Biogeography and Adaptation* and is a frequent contributor to biological journals.

Selected Bibliography

I. General Works

Alderson, L. *The Chance to Survive: Rare Breeds in a Changing World.* Newton Abbot, Devon, England: David and Charles, 1978.

Altevogt, R.F., and B. MacBryde. "Endangered Plant Species," in *McGraw-Hill Yearbook of Science and Technology 1977.* New York: McGraw-Hill, pp. 80-91.

Cadieux, C. *These Are the Endangered.* Washington, D.C.: Stone Wall Press, 1981.

Campbell, F.T. "Conserving Our Wild Plant Heritage." *Environment* 22, no. 9 (November 1980): 14-20.

Council on Environmental Quality. *Biological Diversity.* Reprinted from *The Eleventh Annual Report of the Council on Environmental Quality.* Washington, D.C.: Government Printing Office, 1980.

Council on Environmental Quality. *Global 2000: Entering the 21st Century.* Washington, D.C.: Government Printing Office, 1980.

Curry-Lindahl, K. *Let Them Live.* New York: Morrow, 1972.

Day, D. *The Doomsday Book of Animals: A Natural History of Vanished Species.* New York: Viking Press, 1981.

Eckholm, E. *Disappearing Species: The Social Challenge.* Washington, D.C.: Worldwatch Institute, 1978.

Ehrlich, P.R., and A. Ehrlich. *Extinction: The Causes and Consequences of the Disappearance of Species.* New York: Random House, 1981.

Ehrenfeld, D. *Conserving Life on Earth.* New York: Oxford University Press, 1972.

Elton, C.S. *The Ecology of Invasions by Animals and Plants.* London: Methuen, 1958.

Fisher, J., N. Simon, J. Vincent et al. *Wildlife in Danger.* New York: Viking Press, 1969.

Glacken, C. *Traces on the Rhodian Shore: Nature and Culture in Western Thought from Ancient Times to the End of the Eighteenth Century.* Berkeley: University of California Press, 1967.

Harrison, E.A. *Endangered Species: A Bibliography with Abstracts.* Springfield, Va.: NTIS, 1975.

Harwood, M. "Math of Extinction." *Audubon* 84, no. 6 (November 1982): 18-21.

Hawkes, J.G. *The Diversity of Crop Plants*. Cambridge, Mass.: Harvard University Press, 1983.

Humke, J.W., et al. *Final Report. The Preservation of Natural Diversity: A Survey and Recommendations*. A Report prepared for the U.S. Department of the Interior by the Nature Conservancy, 1975.

International Union for the Conservation of Nature and Natural Resources (IUCN). *Red Data Book*. Morges, Switzerland, IUCN, published in looseleaf form and continually updated. Volume 1, *Mammalia*; Volume 2, *Aves*; Volume 3, *Amphibia and Reptilia*; Volume 4, *Pisces*; Volume 5, *Angiospermae*.

Koopowitz, H., and H. Kaye. *Plant Extinction: A Global Crisis*. Washington, D.C.: Stone Wall Press, 1983.

Linton, R.M. *Terracide*. Boston: Little, Brown, 1970.

Mallinson, J. *The Shadow of Extinction: Europe's Threatened Wild Mammals*. London: Macmillan, 1978.

McCormick, J.F. "Endangered Species—Questions of Science, Ethics, and Law," in *Conference on Endangered Plants in the Southeast, Proceedings*. USDA Forest Service General Technical Report SE-11, 1977, pp. 81-87.

Myers, N. "An Expanded Approach to the Problem of Disappearing Species." *Science* 193 (1976): 198-202.

Myers, N. *The Sinking Ark*. Oxford: Pergamon Press, 1979.

Nilsson, G. *The Endangered Species Handbook*. Washington, D.C.: The Animal Welfare Institute, 1983.

Prance, G., and T.S. Elias. *Extinction Is Forever*. New York: New York Botanical Garden, 1977.

Ripley, S.D., and T.E. Lovejoy. "Threatened and Endangered Species," in P. Mathiesson, ed., *Wildlife in America*. New York: Penguin, 1978.

Shabecoff, P. "New Battles over Endangered Species: Birds and Fish vs. Highways and Dams." *New York Times Magazine*, June 4, 1978, pp. 39-44.

Stewart, D. *From the Edge of Extinction*. New York: Methuen, 1978.

U.S. Congress. *Endangered Species Act Oversight*. Hearings before the Subcommittee on Environmental Pollution of the Committee on Environment and Public Works, U.S. Senate, Dec. 8 and 10, 1981. Washington, D.C.: Government Printing Office, 1982.

U.S. Department of State. *Proceedings of the U.S. Strategy Conference on Biological Diversity* (November 16-18, 1981). Washington, D.C.: Department of State Publication 9262, 1982.

U.S. Department of State. *World Wildlife Conference: Efforts to Save Endangered Species.* Washington, D.C.: Department of State Publication 8729, 1973.

U.S. Fish and Wildlife Service. *Endangered Means There's Still Time.* Washington, D.C.: Government Printing Office, 1981.

Wilson, E.O. *Biophilia.* Cambridge, Mass.: Harvard University Press, 1984.

Wilson, E.O. "The Conservation of Life." *Harvard Magazine,* January-February 1980, pp. 28-37.

Ziswiler, V. *Extinct and Vanishing Animals.* New York: Springer-Verlag, 1967.

II. THE VALUE OF SPECIES

Benn, S.I. "Personal Freedom and Environmental Ethics: The Moral Inequality of Species," in Gray Dorsey, ed., *Equality and Freedom: International and Comparative Jurisprudence,* Vol. 2. Dobbs Ferry, N.Y.: Oceana Publications, 1977, pp. 401-424.

Callicott, J.B. "Animal Liberation: A Triangular Affair." *Environmental Ethics* 2 (1980): 311-338.

Callicott, J.B. "Elements of an Environmental Ethic: Moral Considerability and the Biotic Community." *Environmental Ethics* 1 (1979): 71-81.

Ehrenfeld, D.W. *The Arrogance of Humanism.* New York: Oxford University Press, 1981.

Ehrenfeld, D.W. "The Conservation of Non-Resources." *American Scientist* 64 (1976): 648-656.

Elliot, R. "Why Preserve Species?" in D. Mannison, M. McRobbie, and R. Routley, eds., *Environmental Philosophy.* Australian National University Monograph Series, No. 2, 1980.

Goodpaster, K. "From Egoism to Environmentalism," in K. Goodpaster and K. Sayre, eds., *Ethics and Problems of the 21st Century.* Notre Dame, Ind.: University of Notre Dame Press, 1979.

Gunn, A.S. "Preserving Rare Species," in Tom Regan, ed., *Earthbound: New Introductory Essays in Environmental Ethics.* New York: Random House, 1984.

Gunn, A.S. "Why Should We Care about Rare Species?" *Environmental Ethics* 2 (1980): 17-37.

Kellert, S.R. "Contemporary Values of Wildlife in American Society," in W.W. Shaw and E.H. Zube, eds., *Wildlife Values.* Fort Collins, Colo.: USDA Forest Service Center for Assessment of Noncommodity Natural Resource Values, 1980.

Kellert, S.R. *Public Attitudes toward Critical Wildlife and Natural Habitat Issues. Phase One.* Washington, D.C.: U.S. Fish and Wildlife Service, 1979.

Krieger, M.H. "What's Wrong with Plastic Trees?" *Science* 179 (1973): 446-455.

Leopold, A. *A Sand County Almanac.* London: Oxford University Press, 1949.

Lieberman, G.A. "The Preservation of Ecological Diversity—A Necessity or a Luxury?" *Naturalist* 26 (1975): 24-31.

Norton, B.G. "Environmental Ethics and Nonhuman Rights." *Environmental Ethics* 4 (1982): 17-36.

Passmore, J. *Man's Responsibility for Nature.* New York: Charles Scribner's Sons, 1974.

Raven, P.H. "Ethics and Attitudes," in J.B. Simmons et al., eds., *Conservation of Threatened Plants.* New York: Plenum, 1976.

Rescher, N. "Why Save Endangered Species?" in *Unpopular Essays on Technological Progress.* Pittsburgh: University of Pittsburgh Press, 1980, Chapter 7.

Rodman, J. "The Liberation of Nature." *Inquiry* 20 (1977): 83-131.

Russow, L. "Why Do Species Matter?" *Environmental Ethics* 3 (1981): 101-112.

Sagoff, M. "On Preserving the Natural Environment." *Yale Law Journal* 84 (1974): 205-267.

Singer, P. "Not for Humans Only: The Place of Nonhumans in Environmental Issues," in K. Goodpaster and K. Sayre, eds., *Ethics and Problems of the 21st Century.* Notre Dame, Ind.: University of Notre Dame Press, 1979.

Taylor, P.W. "The Ethics of Respect for Nature." *Environmental Ethics* 3 (1981): 197-218.

Van Dersal, W.R. "Why Living Organisms Should Not Be Exterminated." *Atlantic Naturalist* 27 (1972): 7-10.

Watson, R.A. "Self-Consciousness and the Rights of Nonhuman Animals and Nature." *Environmental Ethics* 1 (1979): 99-129.

III. Endangered Species, Law, and Policy

Burger, W.E. "Majority Opinion." *Tennessee Valley Authority v. Hill,* 437 U.S. Report 153, 173 (1978).

Davis, K.K. *Tellico Dam and Reservoir.* Staff Report to the Endangered Species Committee. Washington, D.C.: U.S. Department of the Interior, 1979.

Erdheim, E. "The Wake of the Snail Darter: Insuring the Effectiveness of Section 7 of the Endangered Species Act." *Ecology Law Quarterly* 9 (1981): 629-682.

Harrington, W. "The Endangered Species Act and the Search for Balance." *Natural Resources Journal* 21 (1981): 71-92.

Harrington, W. "Endangered Species Protection and Water Resource Development." Resources for the Future Working Paper, March 1980.

Harrington, W., and A.C. Fisher. "Endangered Species," in P.R. Portney, ed., *Current Issues in Natural Resource Policy*. Washington, D.C.: Resources for the Future, 1982.

Lovejoy, T.E. "We Must Decide Which Species Will Go Forever." *Smithsonian* 7 (1976): 52-59.

Ramsay, W. "Priorities in Species Preservation." *Environmental Affairs* 5 (1976): 595-616.

Regenstein, L. *The Politics of Extinction: The Shocking Story of the World's Endangered Wildlife*. New York: Macmillan, 1975.

Sagoff, M. "On the Preservation of Species." *Columbia Journal of Law* 7 (1980): 33-67.

Stone, C.D. *Should Trees Have Standing? Toward Legal Rights for Natural Objects*. Los Altos, Calif.: William Kaufman, Inc., 1972.

Stromberg, D.B. "The Endangered Species Act of 1973: Is The Statute Itself Endangered?" *Environmental Affairs* 6 (1978): 511-533.

Travis, M.S. "The Endangered Species Act of 1973." *Harvard Environmental Law Review* 1 (1976): 129-142.

Tribe, L.H. "Ways Not to Think About Plastic Trees: New Foundations for Environmental Law." *Yale Law Journal* 83 (1974): 1315-1348.

U.S. Congress. *Endangered Species Act Amendments of 1982*. Hearing before the Subcommittee on Environmental Pollution of the Committee on Environment and Public Works, U.S. Senate, April 19 and 22, 1982. Washington, D.C.: Government Printing Office, 1982.

Yaffee, S.L. *Prohibitive Policy: Implementing the Federal Endangered Species Act*. Cambridge, Mass.: MIT Press, 1982.

IV. Economics and Endangered Species Policy

Bachmura, F.T. "The Economics of Vanishing Species." *Natural Resources Journal* 11 (1971): 675-692.

Baxter, W.F. *People or Penguins: The Case for Optimal Pollution*. New York: Columbia University Press, 1974.

Bishop, R. "Endangered Species: An Economic Perspective." *Transactions of the 45th North American Wildlife and Natural Resources Conference* (1980), pp. 208-218.

Bishop, R. "Endangered Species and Uncertainty: The Economics of a Safe Minimum Standard." *American Journal of Agricultural Economics* 60 (1978): 10-18.

Ciriacy-Wantrup, S.V. *Resource Conservation: Economics and Politics*. Berkeley and Los Angeles: University of California Division of Agricultural Sciences, 1968.

Clark, C.W. "The Economics of Overexploitation." *Science* 181 (1974): 630-634.

Clark, C.W. *Mathematical Bioeconomics: The Optimal Management of Renewable Resources*. New York: John Wiley and Sons, 1977.

Clark, C.W. "Profit Maximization and the Extinction of Animal Species." *Journal of Politics and Economics* 81 (1973): 950-961.

Fisher, A.C. "Economic Analysis and the Extinction of Species." Report No. ERG-WP-81-4. Berkeley, Calif.: Energy and Resources Group, University of California, November 1981.

Miller, J.R., and F.C. Menz. "Some Economic Considerations in Wildlife Preservation." *Southern Economic Journal* 45 (1979): 718-729.

Myers, N. *A Wealth of Wild Species: Storehouse for Human Welfare*. Boulder, Colo.: Westview Press, 1983.

Pister, E.P. "Endangered Species: Costs and Benefits." *Environmental Ethics* 1 (1979): 341-352.

Ploudre, C. "Conservation of Extinguishable Species." *Natural Resources Journal* 15 (1975): 791-798.

Smith, V.K., and J.V. Krutilla. "Endangered Species, Irreversibility, and Uncertainty: A Comment." *American Journal of Agricultural Economics* 61 (1979): 371-375.

Tucker, W. *Progress and Privilege: America in the Age of Environmentalism*. Garden City, N.Y.: Doubleday, 1982.

V. SCIENTIFIC PRINCIPLES AND ECOLOGICAL MODELS

Brown, D.E., ed. *The Wolf in the Southwest: The Making of an Endangered Species*. Tucson: University of Arizona Press, 1983.

Cody, M.L. "Towards a Theory of Continental Species Diversity," in M.L. Cody and J.M. Diamond, eds., *Ecology and Evolution of Communities*. Cambridge, Mass.: Harvard University Press, 1975.

Farnworth, F.G., and F.B. Golley, eds. *Fragile Ecosystems*. New York: Springer-Verlag, 1974.

Fowler, C.W., and J.A. MacMahon. "Selective Extinction and Speciation: Their Influence on the Structure of Communities and Ecosystems." *American Naturalist* 19 (1982): 480-498.

Frankel, O.H., and M.E. Soule. *Conservation and Evolution*. New York: Cambridge University Press, 1981.

Ghiselin, J. "Wilderness and the Survival of Species: Declining Populations Lose Options in Recessive Genes." *The Living Wilderness*, Winter 1973-74, pp. 22-27.

Hsu, K.J., et al. "Mass Mortality and Its Environmental and Evolutionary Consequences." *Science* 216 (1982): 249-256.

Miller, R.S., and D.B. Botkin. "Endangered Species: Models and Predictions." *American Scientist* 62 (1974): 172-180.

Nitecki, M.H. *Biotic Crises in Ecological and Evolutionary Time*. New York: Academic Press, 1981.

Preston, F.W. "The Commonness and Rarity of Species." *Ecology* 19 (1948): 254-283.

Raup, D.M., and J.J. Sepkoski, Jr. "Mass Extinction in the Marine Fossil Record." *Science* 215 (1982): 1501-1503.

Simberloff, D.S., and L.G. Abele. "Island Biogeography Theory and Conservation Practice." *Science* 191 (1976): 285-286.

Smith, R.L. "Ecological Genesis of Endangered Species: The Philosophy of Preservation." *Annual Review of Ecology and Systematics* 7 (1976): 33-55.

Soule, M.G., and B.A. Wilcox, eds. *Conservation Biology: An Evolutionary-Ecological Perspective*. Sunderland, Mass.: Sinauer Associates, 1980.

Terborgh, J. "Preservation of Natural Diversity: The Problem of Extinction Prone Species." *BioScience* 24 (1974): 715-722.

Vermeij, G.J. *Biogeography and Adaptation*. Cambridge, Mass.: Harvard University Press, 1978.

Whittaker, R.H. "Evolution and Measurement of Species Diversity." *Taxon* 21 (1972): 213-251.

Ziswiler, V. *Extinct and Vanishing Animals: A Biology of Extinction and Survival*. Translated by F. Bunnell and P. Bunnell. New York: Springer-Verlag, 1967.

VI. MANAGEMENT OPTIONS

Diamond, J.M. "The Island Dilemma: Lessons of Modern Biogeographic Studies for the Design of Natural Preserves." *Biological Conservation* 7 (1975): 129-146.

Hoose, P.M. *Building an Ark: Tools for the Preservation of Natural Diversity Through Land Protection*. Covelo, Calif.: Island Press, 1981.

Lovejoy, T.E., and D.C. Oren. "Minimal Critical Size of Ecosystems," paper presented at Symposium on Forest Habitat Islands in Man-Dominated Landscapes. American Institute of Biological Sciences Annual Meeting. East Lansing, Michigan. August 25, 1977.

Holt, S.J., and L.M. Talbot. "New Principles for the Conservation of Wild Living Resources." *Wildlife Monographs*, no. 59 (1978).

Higgs, A.J., and M.B. Usher. "Should Nature Preserves Be Large or Small?" *Nature* 285 (1980): 568-569.

Martin, R.D. *Breeding Endangered Species in Captivity*. New York: Academic Press, 1975.

Olney, P.J.J. "Breeding Endangered Species in Captivity." Zoological Society of London. *International Zoo Yearbook* 17 (1977).

Schonewald-Cox, C.M., et al., eds. *Genetics and Conservation: A Reference for Managing Wild Animal and Plant Populations*. Reading, Mass.: Addison Wesley, 1983.

Terborgh, J.W. "Faunal Equilibria and the Design of Wildlife Preserves," in F. Golley and E. Medina, eds., *Trends in Tropical Ecology*. New York: Academic Press, 1974.

Wemmer, C. "Can Wildlife be Saved in Zoos?" *New Scientist* 75 (1977): 585-587.

Index

Library of Congress Cataloging-in-Publication Data

Main entry under title:

The Preservation of species.
"Written under the auspices of the Center
for Philosophy and Public Policy, University
of Maryland."
Bibliography: p.   Includes index.
1. Wildlife conservation. 2. Species.
3. Extinction (Biology) I. Norton, Bryan G.
II. University of Maryland, College Park.
Center for Philosophy and Public Policy.
QL82.P74 1986   333.95'16   85-42696
ISBN 0-691-08389-4
ISBN 0-691-02415-4 (pbk.)